# The Horsemen of Israel

## HISTORY, ARCHAEOLOGY, AND CULTURE OF THE LEVANT

Edited by

JEFFREY A. BLAKELY *University of Wisconsin, Madison*
K. LAWSON YOUNGER *Trinity Evangelical Divinity School*

1. *The Horsemen of Israel: Horses and Chariotry in Monarchic Israel (Ninth–Eighth Centuries B.C.E.)*, by Deborah O'Daniel Cantrell
2. *Donkeys in the Biblical World: Ceremony and Symbol*, by Kenneth C. Way
3. *The Wilderness Itineraries: Genre, Geography, and the Growth of Torah*, by Angela R. Roskop

# The Horsemen of Israel

## *Horses and Chariotry in Monarchic Israel (Ninth–Eighth Centuries B.C.E.)*

Deborah O'Daniel Cantrell

Winona Lake, Indiana
Eisenbrauns
2011

Printed in the United States of America

www.eisenbrauns.com

**Library of Congress Cataloging-in-Publication Data**

Cantrell, Deborah O'Daniel.
The horsemen of Israel : horses and chariotry in monarchic Israel (ninth–eighth centuries B.C.E.) / by Deborah O'Daniel Cantrell.
p. cm. — (History, archaeology, and culture of the Levant ; 1)
Includes bibliographical references and indexes.
ISBN 978-1-57506-204-4 (hardback : alk. paper)
1. Horses—Israel—History. 2. Chariots—Israel—History. 3. Iron age—Israel. 4. Military history, Ancient. 5. Military art and science—History—To 500. 6. Warfare, Prehistoric—Israel. 7. Bible. O.T.—Criticism, interpretation, etc. 8. Israel—Antiquities. I. Title.
SF285.I75C36 2011
636.109334—dc23
2011018405

The paper used in this publication meets the minimum requirements of the American National Standard for Information Sciences—Permanence of Paper for Printed Library Materials, ANSI Z39.48-1984. ♾™

In Memory of

Mr. and Mrs. Finis A. O'Daniel

*"Perseverance Pays."*

# Contents

# *Preface and Acknowledgments*

When this project began nearly 20 years ago, it was with a simple question and a general curiosity that I naïvely assumed could be quickly and easily answered. Why, I wondered, did Pharaoh's mighty army pursue the fleeing Israelites in chariots? Why didn't the Egyptians ride after them on horseback like cowboys chasing Indians? Didn't people know how to ride "back then"? Thinking the answer might be hidden somewhere in the Old Testament accounts, I began reading slowly and carefully for every reference to equine usage.

Soon, more intriguing questions followed: Why did Joshua hamstring enemy horses and burn their chariots? Did Judge Deborah and Barak fight the Philistines' iron chariots with only an infantry? What exactly were "iron chariots"? Why did King David ride a mule rather than a fine stallion? Why did King Solomon need thousands of horses and chariot cities? How did so many kings of Israel and Judah die in their battle chariots? What was the stimulus for the prophet Elisha's visions of "the horsemen of Israel"?

The search for these answers eventually led me to pursue a Ph.D. in Hebrew Bible at Vanderbilt University, where I was fortunate indeed to meet the brilliant Jack M. Sasson, who opened wide the magical doors of the ancient Near East. Jack ignited the idea for this project by insisting that the ancient world needed a modern-day horse expert. His enthusiasm was contagious, and I found myself traveling to Israel, where I encountered a host of scholars, historians, and archaeologists who wrestled with these same questions. I am honored that they welcomed me to the debate. I am also grateful to Douglas A. Knight for sharing his discerning historical perspective and generously arranging for Vanderbilt students to participate in archaeological digs in Israel. I also appreciate Robert Drews for his keen knowledge of ancient warfare and his willingness to exchange ideas with me.

I am especially indebted to Israel Finkelstein who, after much spirited debate, graciously invited the controversial horses back to Megiddo and who has encouraged every aspect of this endeavor. Thanks also to David Ussishkin, who has always maintained that there were stables in ancient Israel, for helping me realize that scientific "proof" differs from legal "evidence." I am sincerely and forever indebted to my friend and colleague Norma Franklin for kindly welcoming me to Israel, taking me to the numerous sites where history happened, and tirelessly imparting her keen archaeological insights and limitless knowledge.

I appreciate the advice and critique of my editors, K. Lawson Younger and Jeffery A. Blakely, whose input was essential. Any mistakes one may find in this work are all mine; my editors did an excellent job. I gratefully acknowledge the numerous discussions and friendly debates I had with scholars Eric Cline, John Holladay, Aren Maeir, Nadav Naʾaman, Eliezer Piastezky, and Anson Rainey, as well as military historians Richard Gabriel and Steven Weingartner—all of whom contributed significantly to the ideas presented herein. A special thanks to Israeli Olympic qualifier Oded Shimoni, who is truly a modern-day "horseman of Israel," for his steadfast allegiance to the goal of restoring Israel's prominence in the international equestrian world. May you triumph.

My family and friends provided unwavering support and enthusiasm. Oliver, Ryan, Daniel, and Kristin, I thank you for the confidence you happily instilled in this project and for encouraging me to take a hiatus from the practice of law to fulfill this dream. I am grateful to my late father-in-law, Oliver Cantrell, Sr., who believed wholeheartedly in this project and encouraged me with his love and support. Keith O'Daniel provided stalwart wisdom and caustic advice, as only a brother and lawyer can do. I also appreciate Janis Goodwin Moore, whose friendship from the stroller gave her the unqualified right to ask me a million times if I had finished yet. And a special thanks to my friend and neighbor Kathleen Rada Boswell, who dragged me to writing seminars all over the 50 states to improve my style.

Friends come in all sizes and shapes. I am honored to have had some four-legged, hoofed friends with velvet noses and soft, quiet voices. Without them and all they taught me, this mission would have been impossible. So, thank you, Trigger, my first love, and Cicero, Caesar, Buddy (forgiven for breaking my leg), Sundown, Robin Hood, Midnight, Condor, Eager (also known as Dutch Dynamite), Kasteel, Lacot, Royal Copenhagen, Viking, Ali-son, Roz Polo, James Bond, Noway, Glenstern, Dixie Belle, ShowBoat, Fine Time T'nite, Fine China, Fine Print, Wespe, and my current loves, Bling and First Fling. You enriched my life immeasurably. See you in heaven.

Deborah O'Daniel Cantrell, J.D., Ph.D.

# *Abbreviations*

## *General*

| | |
|---|---|
| EA | El-Amarna tablets |
| K. | tablets in the Kouyunjik collection of the British Museum |
| KJV | King James Version of the Bible |
| LXX | Septuagint |
| NIV | New International Version of the Bible |
| NJPSV | New Jewish Publication Society Version of the Bible |
| RSV | Revised Standard Version of the Bible |

## *Reference Works*

| | |
|---|---|
| *ABC* | Grayson, A. K. *Assyrian and Babylonian Chronicles.* Winona Lake, IN: Eisenbrauns, 2000 [original: Locust Valley, NY: Augustin, 1975] |
| *ABD* | Freedman, D. N., et al., eds. *Anchor Bible Dictionary.* 6 vols. New York: Doubleday, 1992 |
| ABRL | Anchor Bible Reference Library |
| *AEL* | Lichtheim, M. *Ancient Egyptian Literature.* 3 vols. Berkeley: University of California Press, 1973–80 |
| *AfO* | *Archiv für Orientforschung* |
| *AJA* | *American Journal of Archaeology* |
| *ANET* | Pritchard, J. B., ed. *Ancient Near Eastern Texts Relating to the Old Testament.* 3rd ed. Princeton: Princeton University Press, 1969 |
| AnOr | Analecta Orientalia |
| *AoF* | *Altorientalische Forschungen* |
| *Arch* | *Archaeology* |
| ARM | Archives Royales de Mari |
| *BA* | *Biblical Archaeologist* |
| *BAR* | *Biblical Archaeology Review* |
| *BASOR* | *Bulletin of the American Schools of Oriental Research* |
| BEATAJ | Beiträge zur Erforschung des alten Testaments und des antiken Judentums |
| BKAT | Biblischer Kommentar: Altes Testament |
| BZAW | Beihefte zur Zeitschrift für die alttestamentliche Wissenschaft |
| *CANE* | Sasson, J. M., ed. *Civilizations of the Ancient Near East.* 4 vols. New York: Scribner, 1995 |
| *CBQ* | *Catholic Biblical Quarterly* |
| CC | Continental Commentaries |
| *COS* | Hallo, W. W., ed. *The Context of Scripture.* 3 vols. Leiden: Brill, 1997–2003 |

| | |
|---|---|
| *ErIsr* | *Eretz-Israel* |
| *HALOT* | Koehler, L.; Baumgartner, W.; and Stamm, J. J. *The Hebrew and Aramaic Lexicon of the Old Testament.* 5 vols. Translated and edited under the supervision of M. E. J. Richardon. Leiden: Brill, 1994–2000 |
| HSS | Harvard Semitic Studies |
| ICC | International Critical Commentary |
| *IEJ* | *Israel Exploration Journal* |
| *JAOS* | *Journal of the American Oriental Society* |
| *JARCE* | *Journal of the American Research Center in Egypt* |
| *JBL* | *Journal of Biblical Literature* |
| *JCS* | *Journal of Cuneiform Studies* |
| *JHS* | *Journal of Hebrew Scripture* |
| *JJS* | *Journal of Jewish Studies* |
| *JNES* | *Journal of Near Eastern Studies* |
| *JQR* | *Jewish Quarterly Review* |
| *JSOT* | *Journal for the Study of the Old Testament* |
| *JSS* | *Journal of Semitic Studies* |
| LAPO | Littératures anciennes du Proche-Orient |
| LCL | Loeb Classical Library |
| *MHQ* | *Military History Quarterly* |
| *NEAEHL* | Stern, E., ed. *The New Encyclopedia of Archaeological Excavations in the Holy Land.* 4 vols. Jerusalem: Israel Exploration Society and Carta / New York: Simon & Schuster, 1993 |
| OBO | Orbis biblicus et orientalis |
| *OEANE* | Meyers, E. M., ed. *The Oxford Encyclopedia of Archaeology in the Near East.* 5 vols. New York: Oxford University Press, 1997 |
| OIC | Oriental Institute Communications |
| OIP | Oriental Institute Publications |
| OLA | Orientalia lovaniensia analecta |
| OtSt | Oudtestamentische Studiën |
| *PEQ* | *Palestine Exploration Quarterly* |
| *RA* | *Revue d'assyriologie et d'archaéologie orientale* |
| *RlA* | Ebeling, E., et al., eds. *Reallexikon der Assyriologie.* Berlin: de Gruyter, 1924–32 |
| SAA | State Archives of Assyria |
| *SAAB* | *State Archives of Assyria Bulletin* |
| SAAS | State Archives of Assyria Studies |
| SBLWAW | Society of Biblical Literature Writings from the Ancient World |
| *TA* | *Tel Aviv* |
| *TFS* | Dalley, S., and Postgate, J. N. *The Tablets from Fort Shalmaneser.* Cuneiform Texts from Nimrud 3. London: British School of Archaeology in Iraq, 1994 |
| *UF* | *Ugarit-Forschungen* |
| *VT* | *Vetus Tetamentum* |
| *ZDPV* | *Zeitschrift des Deutschen Palästina-Vereins* |

## Chapter 1

# *Introduction*

*My, father! My father! The chariots and horsemen of Israel!*
*(2 Kgs 2:12, NIV)*

Restoring horses to their proper place in the historical landscape of Iron Age Israel enriches the ancient narrative in various surprising and interesting ways. The quotation above cites the prophet Elisha as he witnesses the fiery chariots of God swooping from the clouds to translate Elijah to heaven. Horses, chariots, and the "horsemen of Israel" had become so prevalent during the Monarchic period that they fired the poetic imagination of prophets and inspired the poets. Their participation in the life and culture of biblical Israel became a deeply entrenched memory in the collective consciousness. Almost every book in the Hebrew Bible involves horses and chariots in some manner, usually in a military context.

Yet, the importance of horses, chariots, and equestrians in ancient Israel typically is mentioned only in passing, if at all, by historians, hippologists, and biblical scholars. When cited, the topic engenders much confusion. It is odd that, for the most part, the history of Israel is written without reference to the practical realities that a state-run, horse-management program entails. Large numbers of horses require a sophisticated, governmental infrastructure to support them. At a minimum, such an organization taps into a substantial labor force, determines agricultural policies, and dictates architectural accommodations designed specifically for horses. An administrative structure of this sort necessarily affects the social order and, for at least this reason, the subject merits examination.

Horses leave no agricultural or architectural imprints on their habitation. They eat only grass and grain, and their dung naturally fertilizes the soil. They drink from streams and rivers and build no burrows, nests, or dens. When they die from injury or illness (speed giving them advantage over most predators), their carcasses are picked clean by scavengers; their bones are scattered. However, once humans become involved in horse management—whether for national security and military conquest, as in the ancient world, or simply for transportation or profit—they invest a great deal of time and effort in training and building the attendant architectural accommodations to maximize the horse's potential. It is then that archaeologically we find stables, large courtyards, hitching stations, and traces of training accoutrements.

### Background

Notwithstanding the substantial textual and archaeological evidence of the horse's historic presence, recent scholarship seems led by a general belief that there were very few horses in Iron Age Israel and that Israel's chariotry was insignificant. The reason for this current sentiment is tied primarily to the academic controversy over the past 50 years about whether the 17 tripartite pillared buildings excavated at Megiddo in the early twentieth century were, in fact, stables. Although the original excavators, archaeologists from the University of Chicago, designated these buildings (some with feeding troughs in situ) as stables, a number of scholars (and a few archaeologists) later challenged this view and adopted alternative interpretations. After they "reassessed" the Megiddo stables as "storehouses," "marketplaces," or "barracks," the idea developed that there was no place for the horses to live and, therefore, there must have been few horses in Israel. The lack of stables, when added to the suggestion that Iron Age Israel could not afford to buy expensive horses and maintain an even-more-expensive chariotry, led to a dearth of horses in ancient Israel; or so the logic goes that has permeated the scholarship. This book attempts to dispel that notion.

### Modern Horsemanship

In today's mechanized society, intimate knowledge of horses is rare, and therefore, much misunderstanding can enter into interpretation of any ancient work that cites them. Moreover, our modern viewpoint of horses is skewed somewhat by the way the thoroughbred racing sport uses two- and three-year-old horses, retiring them by age five, a time when most horses are just beginning their careers. Thoroughbred racing, however, is its own peculiar discipline dedicated totally to speed, with no emphasis on the collection, control, and restraint required for chariotry or cavalry maneuvers. Although racing is the horse sport most accessible to the general public, it is decidedly unrelated to the practical requirements of ancient chariotry and cavalry. Likewise, a dude-ranch horse used for trail riding, familiar to many from vacation rides, or a hack horse providing carriage rides for city tourists bears little or no resemblance to the highly trained, intelligent, skilled athletes that were required to perform as war-horses in antiquity.

The most relevant modern equine sport comparisons to the training of ancient war-horses are found in the modern disciplines of dressage, an Olympic sport involving classical cavalry maneuvers; and two- or four-in-hand driving, a carriage-maneuvering sport recognized by the Fédération Equestre Internationale (FEI) in 1970. My involvement with horses as a rider, competitor, trainer, breeder, and importer includes equine experience ranging from competitive barrel-racing to jumping, and for the past 25 years, dressage. This study relies on the knowledge I have obtained from exposure to several breeds of horses: English and American thoroughbreds, American quarter horses, Arabians, Tennessee walking horses, and European warmbloods. My respect and admiration for the animal, whatever its breed, is undiminished.

However, I believe that, because ancient societies relied on horses for their military and economic needs, rather than for pleasure, they developed ways of training and caring for them that may surpass what is the norm today. Likewise, it would be inappropriate to assume that modern horsemanship surpasses that of riders in antiquity, if for no other reason than the fact that today's equestrians do not depend on their horses to save their lives in battle. In the ancient world, a well-trained and well-conditioned horse made the difference between life and death. To know the nature of the horse taking a soldier to battle as well as its limitations was crucial.

## *Ancient Scholarship*

In some ways, the knowledge of horse care and training was (and still is) its own international discipline based on thousands of years of accumulated experience. This premise is confirmed by some of the most ancient documents yet discovered that involve the training and care of horses. For example, the Kikkuli text, dating to the fourteenth century B.C.E., is a training manual for the conditioning of chariot horses found at the Hittite capital, Hattusa.[1] The master horse trainer, Kikkuli, recommends a six-month regimen that includes stabling, exercise, grooming, swimming, and special diets. Much of the Kikkuli protocol, which includes interval training, is still valid today, especially for the sport of endurance riding. Other ancient works, such as the Hippiatric texts from Ugarit (fourteenth century) and Xenophon's cavalry manual and treatise on horsemanship (fourth century) also contain timeless information about horses. Even though *The Iliad* is not primarily historical, it likely reflects the realities of the periods of its compilation and might therefore give useful information about horses and chariotry. Scholars have frequently consulted these ancient works and others, making them foundational for this study.

## *Current Scholarship*

Modern hippological authors and historians, such as Robert Drews, Mary Littauer, and Ann Hyland write with great insight about the use of horses and chariotry in the ancient world. Although their works delve into horsemanship in Egypt, Greece, and Assyria, not much focus is given to Israel, probably because biblical timelines can be very difficult to establish, and the historicity of the narratives is notoriously

1. All dates are B.C.E., unless otherwise designated.

difficult to assess. Assyriologists Stephanie Dalley, John Postgate, Frederick Fales, and more recently, Tamás Dezsö have studied the structure and composition of the Assyrian army, necessarily touching on its interaction with Israel's chariotry, especially after the fall of Samaria, when many of the horses and Israelite/Samarian chariot officers were incorporated into Assyria's home army. Their work, however, does not extend to the detailed operation of an Iron Age horse-management program inside Israel proper.

A recent work, *The Military History of Ancient Israel*, by Richard Gabriel, dissects the battle strategies and combat techniques as described in the biblical texts from the exodus through Solomon. While Gabriel frequently mentions chariotry, his focus is primarily on its advantages and disadvantages in particular battles and not on the actual role of the horses. His work is most helpful as a reference for the logistics and strategic thinking necessary in waging war. However, the scope of Gabriel's work ends where this book begins.

Yigael Yadin's classic work *The Art of Warfare in Biblical Lands* assumes a large and vibrant chariotry in Iron Age Israel based on archaeology and "confirmed" by the biblical text. Yadin's treatise, although somewhat outdated contains an extensive collection of photos and plates depicting ancient Egyptian and Assyrian warfare; I cite them frequently for convenience. Also, I include various images of Assyrian horses and chariots from the palace reliefs currently on display at the British Museum. I selected images that illustrate a specific point, usually about warfare or riding. I am not suggesting that because we see beautiful, sleek, well-groomed, richly caparisoned horses posed with the kings of Assyria that these horses were the usual army mounts. Undoubtedly the regular army rode mounts less majestically outfitted.

Some Israeli historians and scholars such as Nadav Na'aman, Anson Rainey, and Moshe Elat have reached tentative conclusions against the presence of large numbers of horses in Israel for various reasons. Other scholars have followed their lead, questioning whether Israel had room for many chariots, and where all the stables would be for more than 4,000 horses.[2] In general, their failure to consider the nature of horses and to recognize the basic components of their care distorts their conclusions. However, based on the extensive evidence at Megiddo and indications at other sites such as Jezreel and Lachish, Israeli archaeologists Israel Finkelstein, David Ussishkin, and Norma Franklin acknowledge the value of horses in Iron Age Israel and are convinced of their role in warfare and trade.

2. See, for example, G. W. Ahlström, "The Battle at Ramoth-Gilead in 841 B.C.," in "*Wünschet Jerusalem Frieden*": *Collected Communications to the XIIth Congress of the International Organization for the Study of the Old Testament* (BEATAJ 13; Frankfurt am Main: Peter Lang, 1988) 157–66 n. 4.

## *Historical Setting*

The history and archaeology of Megiddo are central to the discussion. Located at the crossroads of the land bridge between Egypt and Mesopotamia, Megiddo's geopolitical importance was unsurpassed in the ancient world. From the third millennium on, imperial powers sought to control Megiddo for its lucrative trade routes and the strategic military advantage it afforded. Megiddo's involvement with horses and chariots begins during the Late Bronze Age (1550–1150), because one of the first recorded battles in history was fought there in 1479, between the Egyptians and a Canaanite coalition.[3]

The historical setting of this book, however, involves primarily Iron Age Israel's connection with horses after the invasion of Shoshenq I (Shishak, ca. 925) through the capture of Megiddo by the Assyrians in 732, and in Judah through the siege of Jerusalem in 701 to the death of Josiah in 609. It was during the roughly 200 years from 925 to 732 that Israel (the Northern Kingdom) and to a lesser extent, Judah (the Southern Kingdom) used their respective horses and chariotries to protect their borders, establish their statehood, and on occasion to invade and conquer more territory. The horses and chariots headquartered at Megiddo and Jezreel were critical to the survival and continued existence of both kingdoms. Quite simply, without the chariots and horsemen of Israel, there would have been no Israel (as we know it), at least during the Iron Age.

The use of the term *Iron Age* in this work refers to both Iron Age I (1150–900) and Iron Age II (900–586). This time period was chosen because it covers roughly the historical settings in the Hebrew Bible from Saul's death while fleeing the Philistine chariot archers in the Jezreel Valley (ca. 1000) to the death of Josiah in a chariot battle against Pharaoh Necho at Megiddo (ca. 609). Whether these or any other specific Hebrew Bible accounts are legends, well-preserved memories, or historical records is not critical for the purposes of this study. It is the fact that the traditions mentioning horses survived and continue to exist that compels further inquiry and investigation. The chronology of events is far less important than the verisimilitude with which the horses inform the texts. It is not essential for this study that the Hebrew Bible deal with real-time events, because the basic nature of horses and their needs and requirements have not changed for 3,000 years. The mission of this work is to consider what impact horses had on the history and culture of Israel and Judah, whenever and wherever they were present during the Iron Age.

Scholars differ on the dates attached to the kings of Israel and Judah. Fortunately, in this book there are few contexts in which the discrepancy in dating creates

3. R. Krauss, "Lunar Days, Lunar Months, and the Question of the Civil-Based Lunar Calendar," in *Ancient Egyptian Chronology* (ed. E. Hornung, R. Krauss, and D. A. Warburton; Handbook of Oriental Studies, Section 1: The Near and Middle East; Leiden: Brill, 2006) 395–431.

a problem. The dates I cite come from the 2006 edition of *Carta's Atlas of the Biblical World* by Anson F. Rainey and R. Steven Notley.[4] Readers may also want to consult Mordecai Cogan's article, "Chronology" in the *Anchor Bible Dictionary,* as well as the tables compiled in John Rogerson, *Understanding the Old Testament: Chronicle of the Old Testament Kings.*[5]

In recalling Israel's premonarchic history, the biblical authors associate horses with established enemy forces such as Egypt, Philistia, Canaan, and Aram (see chap. 6), indicating that horses were associated with governmental/urban power. The notion of statehood for Israel and Judah during the Iron Age necessarily required the support of horses, chariotry, and a standing army as basic survival tools, because the neighboring countries had these valuable weapons at their disposal. The necessity to train the horses to function effectively as military units required the organizational skills and administration of a state-run entity to build stables and training compounds. The much less trustworthy alternative would be reliance on mercenary forces, which happened on occasion.

It is fair to assume that all references to horses in the Hebrew Bible were first written (and later edited) by persons close in time and experience to the military use of horses in battle. This assumption holds true for the entire Iron Age setting of the various battle narratives and also for the later historical setting of the authors, whether from the Pre- or Postexilic period. For example, although scholars typically place the authorship of Chronicles during the Persian period in the latter half of the fifth century, with some authorities believing it to be even later, during the Hellenistic period, the book of Chronicles is still useful for this work.

Even though scholars are reluctant to rely on the Chronicles for historical information because they were written some four centuries after the formation of Israel's statehood and presumably by authors with a nationalistic agenda, there is no reason to ignore what is written about horses and their role in battle simply because it appears only in Chronicles. In fact, by the Persian period (or Hellenistic period), the world had more, not less, collective experience about the nature of the horse and its importance in battle. In addition, the writers during the Persian period and later were writing after the demise of chariot warfare and the rise of mounted combat. Therefore, when the Chronicler mentions an ancient chariot battle, it may mean that the tradition about that battle survived (by oral tradition or in military/historical documents no longer available) long past the time when chariots were key to warfare. These accounts therefore merit consideration and critical inquiry.

4. A. F. Rainey and R. S. Notley, *The Sacred Bridge: Carta's Atlas of the Biblical World* (Jerusalem: Carta, 2006).

5. M. Cogan, "Chronology," *ABD* 1:1002–11; J. W. Rogerson, *Chronicle of the Old Testament Kings: The Reign-by-Reign Record of the Rulers of Ancient Israel* (London: Thames & Hudson, 1999). For further discussion of these issues, see I. Finkelstein and N. A. Silberman, *David and Solomon: In Search of the Bible's Sacred Kings and the Roots of the Western Tradition* (New York: Free Press, 2006) 17–20.

Of course, the issue of whether a battle actually occurred as detailed in Chronicles or any other book, whether from the Hebrew Bible or Assyrian, Egyptian, or Greek battle accounts can be vexing because military rhetoric sometimes diverges from accurate reporting. And, as is true of all ancient writings, the history is devilishly hard to disentangle from the fiction. While it is important to acknowledge this fact, it is equally important to persevere in dealing with the Hebrew Bible texts as at least verisimilar. Therefore, although for narrative purposes I may present a particular biblical battle cinematically, whether the battle actually occurred or not (when, where, and against whom) is often not the main focus of this book. Rather, I try to evaluate each account for its capacity to give horse and chariot their proper places in the military life of Israel. From this perspective, whether true to history or not, Hebrew narrators often prove remarkably realistic, and their contributions worthy of giving readers dependable chapters on major moments in Israel's cultural history.

It is now almost a century since horses played a significant role in modern warfare; nearly all accounts of battle before 1918 C.E. involved horses to some extent, and their usefulness is typically acknowledged in the scholarship of the time. Too often today, however, scholars ignore or diminish the role of the horse in battle. It is important to remember that ancient historians took for granted knowledge about horses that modern scholars have now forgotten or never knew.

## *Biblical Text*

One problem encountered by any scholar using the biblical text for reference is which translation to select for quotations. For this book, the Tanakh translation from the Jewish Publication Society (NJPSV) is used as the primary resource. However, the NJPSV often employs the annoying convention of referring to horses as "steeds," a term that may have been popular in medieval terminology but has no meaning whatsoever in the modern equestrian world. In fact, applying the word *steed* to a horse is somewhat akin to using the word *mammal* for a human; it is correct but minimally so. Therefore, to avoid the *steed* terminology, I sometimes select alternative translations from the New International Version (NIV), the Septuagint (LXX), or for an occasional poetic rendering, the King James Version (KJV), and I indicate the version in the text.

Another significant challenge in deciphering the biblical references involves the Hebrew designations for "horse" and "horsemen." The typical Hebrew words used for 'horse' *sûs* (סוס) and for 'chariot,' *rekeb* (רכב) are simple and straightforward. It is the Hebrew word *pārāš* (פרש), which can be translated either 'horse' or 'horseman', that causes translators much difficulty. In some instances, there is not a great difference between the meaning of 'horse' and 'horseman'; presumably, 'horseman' could indicate either a cavalryman or charioteer. When encountering this word in the text, we must consider the context of the narrative to determine which word is more suitable.

This issue typically arises when the narrative involves warfare: the translation of *pārāšîm* (פרשים) as 'horsemen' could convey the idea of cavalry; whereas, the translation 'horses' might convey the notion of chariotry.[6] The practical difference is that a cavalryman rides one horse, and chariotry involves at least two or three horses. In this study, for all direct quotations of biblical material using *pārāšîm*, I have considered the context and selected the translation that best conveys the appropriate meaning.[7]

### Chronology Matters

The ongoing archaeological debate concerning the low chronology of Iron Age Israel, which involves about an 80–100-year difference in the dating of archaeological strata between the eleventh to ninth centuries, affects this work.[8] Regarding the Megiddo stables, the low chronology places their construction around 800 and, therefore, their usage during the reign of Jeroboam II (801–786) and his successors until the invasion of Tiglath-pileser in 732. The historical setting of Jeroboam's expansion activities and the probability that Megiddo was used as a major sales center for trained chariot horses fits well within the low chronology timetable.[9]

However, if the Megiddo stables were constructed in the early ninth century, as assumed by Yadin and others, they may have served as Ahab's military headquarters, horse compound, and mustering center prior to the Battle of Qarqar. As explained by Israel Finkelstein in *Megiddo IV*, the archaeological evidence does not suggest the presence of elaborate stables during the Omride period.[10] This does not mean that there were no horses at Megiddo; there assuredly were. It simply means they were

6. For the suggestion that *pārāš* is a synonym for *sûs* and should be translated 'chariot horse' when used in conjunction with *rkbl mrkbh* as found in 1 Sam 8:11, 2 Sam 15:1, and 1 Kgs 1:5, see L. E. Stager, "Chariot Fittings from Philistine Ashkelon," in *Confronting the Past: Archaeological and Historical Essays on Ancient Israel in Honor of William G. Dever* (ed. S. Gitin, J. E. Wright, and J. P. Dessel; Winona Lake, IN: Eisenbrauns, 2006) 169–76.

7. For a detailed analysis of the etymology and the various considerations used by translators in attempts to arrive at a solution to the long confused usage of *pārāšîm,* see H. Niehr, "פרשׁ," *TDOT* 12:124–28. For a history of the translation problems with *pārāšîm*, see G. I. Davies, "ʾURWŌT in I Kings 5:6 (EVV. 4:26) and the Assyrian Horse Lists," *JSS* 34 (1989) 25–38. For the suggestion that almost every use of *pārāšîm* means 'chariot horses' rather than 'horsemen', see S. Mowinckel, "Drive and/or Ride in O.T.," *VT* 12 (1962) 278–99.

8. For a recent discussion of these issues and the scholarly divide, see A. Mazar, "The Spade and the Text: The Interaction between Archaeology and Israelite History Relating to the Tenth–Ninth Centuries B.C.E.," in *Understanding the History of Ancient Israel* (ed. H. G. M. Williamson; Proceedings of the British Academy 143; Oxford: Oxford University Press, 2007) 143–71.

9. The concept of Megiddo as a major sales center was discussed by N. Franklin, "Fair or Foal" (Paper resented at the annual meeting of the Society of Biblical Liberature, Toronto, 23 November 2002). See also H. Shanks, "Horsing Around in Toronto," *BAR* 29/2 (2003) 50–53.

10. D. O. Cantrell and I. Finkelstein, "A Kingdom for a Horse: The Megiddo Stables and Eighth Century Israel," in Megiddo *IV: The 1998–2000* (ed. I. Finkelstein, D. Ussishkin, and B. Halpern; 2 vols; Monograph Series 24; Tel Aviv: Tel Aviv Institute of Archaeology, 2006) 2:643–65.

stabled elsewhere on or near the compound, perhaps in wooden structures or brush arbors.

Regarding other archaeological sites in Israel, including sites with stables and/or four- and six-chambered gates, whether the construction dates vary by 100 years is not at issue in this book. It is noteworthy that stables dating to the tenth and ninth centuries have been identified by archaeologists at numerous sites in Israel, Judah, Philistia, and Phoenicia. In fact, the expansive Megiddo stabling and training compound is emphasized in this work because it represents the apex of a long history of the development of chariotry and riding and the complimentary architectural structures to support equine pursuits. Certainly the stables and training compound at Lachish in Judah dating to the late tenth–early ninth century are equally as important as the headquarters of the Judean chariotry and cavalry forces. Further study of the interconnectedness of the fortresses, stables, and gates may contribute to the chronology debate but is beyond the scope of this book. For purposes of this work, the significant point is that the stables and chambered gates were *present* in the Iron Age and were designed and used for horse-management purposes, whenever they were constructed.

## *Horse-Related Architecture*

Horses do not require stables to thrive. However, horse trainers from Kikkuli to the present know that horses learn faster and more reliably when they experience the routine of stabling, regular feeding, grooming, and consistent human contact as part of a regimented program. Once trained, horses do not need to remain stabled; they may be kept in corrals, pens, or enclosed courtyards. Therefore, it seems certain that the occurrence of large stables was specifically for horses in training for economic or military purposes. This means that training facilities, such as Megiddo, with stables for 450 horses; the Jezreel fortress with its huge enclosed courtyard; and Lachish with its courtyard, chambered gates, and stables were in fact processing thousands of horses, only some of which required stabling at any given time. In addition, mares and foals, having very different needs, were probably kept in pasture areas away from the training complexes. There is no accurate way to estimate the number of horses in Iron Age Israel and Judah, but the available evidence suggests that several thousand horses would be expected in Northern Israel alone.

Stables or shelters were especially important to protect the horses from the sun. Unlike sheep, goats, donkeys, cattle, and camels, which live year round without the need for shelter, horses are susceptible to sunburn.[11] Knowledge of horses'

11. For this reason, the 25,000 horses in General Allenby's cavalry stationed in the Jordan Valley in the summer of 1918 C.E. wore blankets to protect their skin when shelter was not available. Marquess of Anglesey, *A History of the British Cavalry, 1816–1919*, vol. 5: *A History of the British Cavalry, 1914–1919 Egypt, Palestine and Syria* (London: Pen and Sword, 1994) 233.

comparatively sensitive skin suggests that buildings similar to today's "run-in" sheds may have been erected for their shelter during the Iron Age. Obvious candidates for these horse shelters are the controversial tripartite pillared buildings found at numerous sites throughout Israel and Judah, which were usually constructed conveniently near the compound entrance. These buildings should be reevaluated for their potential horse usage during the Iron Age, especially because simultaneous usage as storehouses was easily possible.

Additionally, the four- and six-chambered gates are unique, architectural marvels peculiar to the Levant. It is more than coincidence that their measurements match those of horses and chariots and that all new construction on them disappears by the seventh century with the demise of chariotry as the preferred warfare convention. Their strategic locations need further study to understand better the role that they may have played in an interconnected defensive network of fortresses in Israel and Judah. The chambered gates, used as convenient hitching stations, may identify the "chariot cities" of Israel. If so, the debate about who built them and when should be reignited in view of recent archaeological studies.

### *Contents*

This book begins with the biology of the war-horse because, unless we understand something about the nature of the horse, we cannot understand its true significance in Iron Age warfare. Chapter 3 discusses the textual and archaeological evidence for the presence of horses in Israel and provides theories about how large numbers of horses may have been acquired. Chapter 4 considers the strategic operation of chariotry forces in Israel and the architectural support for them. The academic controversy over the stables at Megiddo is addressed in chapter 5, as well as the practical operations of a large training center and horse compound there and at Jezreel. Chapter 6 comments on the use of horses in warfare in Iron Age Israel and the equine implications of various enemy invasions as set forth in the biblical narrative. Chapter 7 describes the evolution of warfare convention from chariotry to mounted combat that occurred during the Iron Age.

I hope that this study, which concentrates primarily on the horse and how the ancient world answered its needs, will be helpful in bridging the historical gap in scholarship, which generally omits Israel's equestrian accomplishments and accommodations. It is only a beginning. The primary goal is to spark informed debate and further investigation of this fascinating topic.

## Chapter 2

# *The Nature of the War-Horse*

*Horses are a false hope for deliverance;*
*for all their great power they provide no escape.*
*(Ps 33:17)*

Addiction to speed is the single most salient quality of a war-horse. The battle horse shivers, trembles, leaps, stomps, snorts, tosses its head, and sometimes rears up and paws the air in anticipation of running all out, into the frenzy of battle.[1] The exhilaration of the moment creates a sensory overload, causing the horse to dance, prance, and quiver from head to hoof as it tries to break loose from the restraint of human control to reach bursts of speed over 40 miles per hour.[2] Today, this sort of anticipatory excitement radiates as a nervous racehorse dances to the starting gate, as a polo pony leaps and bobs in place on the sidelines just before taking the field, or as a barrel racer at the rodeo thunders into the arena.[3] In the ancient world, the awe-inspiring energy and power of the horse was most often witnessed in battle.

1. For a detailed discussion of the mechanics of movement in a horse, see Stephen Budiansky, *The Nature of Horses: Exploring Equine Evolution, Intelligence, and Behavior* (New York: Free Press, 1997) 175–210.

2. Thoroughbred racehorses, polo ponies, and harness-racing horses pulling a 35-lb. sulky (cart) and driver can attain speeds of 45 mph over short distances. D. Ours, *Man O'War: A Legend like Lightning* (New York: St. Martins, 2006) 158–61. Horses alter the speed of a gait by adjusting tempo and stride length. There are distinct changes in speed among the various types of canter: collected canter, 7.2 mph; working canter, 8.7 mph; medium canter, 10.9 mph; and extended canter, 13.2 mph. Horses are taught to use these gaits on command after months and years of training. H. M. Clayton, "The Canter Considered," *USDF Connection* 4/6 (2002) 22–25. Some experts believe that the lightweight Bronze Age chariots (75 lbs.) attained maximum speeds of only 20–30 mph. See J. Keegan, *A History of Warfare* (New York: Knopf, 1994) 159; and W. J. Hamblin, *Warfare in the Ancient Near East to 1600 B.C.: Holy Warriors at the Dawn of History* (London: Routledge, 2006) 145. Based on personal experience, I believe these estimates are conservative and that horses, if required, could easily pull chariots at full speed over short distances on the battlefield. The main problem was loading the chariots with enough weight to keep them grounded, as discussed in chap. 4 below.

3. Barrel racing is a rodeo event in which the competitor rides the horse around three oil-drum-size barrels arranged in an equidistant triangle. The horse with the fastest time wins. J. Mayo and B. Gray, *Championship Barrel Racing* (Cypress, TX: Cordovan, 1961) 9–11.

The basic nature of the horse, as is true of any prey animal, is to flee danger, not run toward it. Nevertheless, as a seasoned war-horse neared the battlefield, its speed addiction subverted its natural instinct to flee. Rather than flee, the war-horse strove to *explode* onto the battle scene, much as the racehorse seeks to burst from the gate at the racetrack. The war-horse's desire to charge into battle symbiotically melded with its internal flight mechanism, and a new energy emerged, a spirit of courage against all odds. It is this animated, charging war-horse that is frequently chronicled in ancient texts, particularly in the Hebrew Bible: "Charge, O horses! Drive furiously, O charioteers!" (Jer 46:9); "He paws with force, he runs with vigor, *charging into battle*" (Job 39:21; emphasis added); "They all persist in their wayward course like a horse *dashing forward in the fray*" (Jer 8:6; emphasis added); "*galloping horse* and bounding chariot! Charging horsemen, flashing swords, and glittering spears!" (Nah 3:2–3; emphasis added).[4]

Without the proper training, the war-horse was a dangerous liability on the battlefield.[5] The paradoxical tendencies of the horse, charging and fleeing, had to be curbed and trained before it was moderately safe and useful in battle.[6] For charioteers and cavaliers, the war-horse was a trained partner, much stronger than they but totally dependent on them for decision-making; each could save the other's life and no doubt often did. As eloquently expressed by the fifth-century Greek cavalryman, Xenophon (444–357), "for it is plain that in danger the master entrusts his life to his horse."[7] Indeed, Ramesses II (1279–1213) credited his chariot horses Mut-Is-Content and Victory-in-Thebes with saving his life at the Battle of Qadesh (1285):

4. See also "His majesty (Ramses II) then rushed forward; at a *gallop he charged the midst of the foe*. For the sixth time he *charged* them" (M. Lichtheim, "The Kadesh Battle Inscriptions of Ramses II," *Ancient Egyptian Literature* [3 vols.; henceforth abbrev. *AEL*; Berkeley: University of California Press, 1973–80] 2:69). "Hector's horses were *charging out to battle*, galloping full stretch" (Homer, *The Iliad* [trans. Robert Fagles; New York: Penguin, 1991] 16.439.972–73). In the provocative conversation between Gilgamesh and Lady Ishtar in "The Bull of Heaven," he reminds her, "You loved the horse, so famed in battle, but you made his destiny whip, spur and lash" (A. George, *The Epic of Gilgamesh: A New Translation* [New York: Barnes & Noble, 1999] IV 53–54, p. 49).

5. Horse training was already an established science in Mitanni and Hatti by the mid-fourteenth century. Kikkuli, a Mitanni master horse trainer living in Hattusa, wrote a detailed treatise on training techniques for chariot horses that specified interval training (often at night), swimming in rivers, drinking beer, saunas, and full-body massages. Ann Nyland developed a modern-day training program for endurance horses based on Kikkuli training techniques (*The Kikkuli Method of Horse Training*, http://www.kikkulimethod.com [accessed 21 May 2006]).

6. An overly excited war-horse is credited with contributing to the death of Cyrus the Younger in 401 at the Battle of Kounaxa. For two versions of this incident, see Plutarch, *Artaxerxes*, available on the Internet Classics Archive, http://classics.mit.edu/index.html. See also R. Drews, *Early Riders: The Beginnings of Mounted Warfare in Asia and Europe* (New York: Routledge, 2004) 142–43.

7. Xenophon, *On the Art of Horsemanship* in *Scripta Minora* (trans. E. C. Marchant and G. W. Bowersock; vol. 7; LCL 183; Cambridge: Harvard University Press, 2000) 4.1.

I crushed a million countries by myself
On Victory-in-Thebes, Mut-is-Content, my great horses;
It was they whom I found supporting me,
When I alone fought many lands.

They shall henceforth be fed in my presence,
Whenever I reside in my palace.[8]

The adrenalin-infused war-horse charging into the fury and turmoil of battle made a lasting impression on ancient poets and writers.[9] The imagery is captured vividly in Job 39:19–25:[10]

Do you give the horse his strength?
Do you clothe his neck with a mane?

Do you make him quiver like locusts,
His majestic snorting [spreading] terror?

He paws with force, he runs with vigor,
Charging into battle.

He scoffs at fear; he cannot be frightened;
He does not recoil from the sword.

A quiverful of arrows whizzes by him,
And the flashing spear and the javelin.

Trembling with excitement, he swallows the land;
He does not turn aside at the blast of the trumpet.

As the trumpet sounds, he says, "Aha!"
From afar he smells the battle,
The roaring and shouting of the officers.

The above passage is set forth in detail because it provides an interesting framework for discussing the nature of the war-horse. If, as has been suggested, this passage dates to the sixth century, it may be one of the oldest and most detailed literary descriptions of a horse anticipating battle.[11] Whether the Job 39 author envisioned a chariot horse

8. Lichtheim, "The Kadesh Battle Inscriptions of Ramses" (*AEL* 2:62–72). For an insightful description of the chariotry clashes in the Battle of Qadesh from an equestrian's viewpoint, see A. Hyland, *The Horse in the Ancient World* (Westport, CT: Praeger, 2003) 85–88.

9. "Indeed a prancing horse is a thing so graceful, terrible and astonishing that it rivets the gaze of all beholders, young and old alike" (Xenophon, *Art of Horsemanship* 11.9).

10. For a literary analysis of the Job 39 war-horse as a paradigm for inner violence, chaos, and undeserved suffering, see D. Odell, "Images of Violence in the Horse in Job 39:18–25," *Prooftexts* 13 (1993) 163–73.

11. Other ancient texts acknowledge the high status of the war-horse: for example, the delightful Assyrian fable (seventh century), "The Ox and the Horse," in which the two creatures debate which of them is the strongest, bravest, and most beneficial to society (W. G. Lambert, *Babylonian Wisdom Literature* [Oxford: Clarendon, 1960; repr. Winona Lake, IN: Eisenbrauns, 1996] 175–83).

Fig. 2.1. Head of Herakles wearing lion-skin face mask. Alexander III struck ca. 320–317, AR Tetradrachm, Macedonia. Author's collection.

or a mounted charger is neither clear nor essential to the accuracy of the discourse; either horse would react with the same intensity to the sights, sounds, and smells associated with the battlefield. I envision a chariot horse, perhaps the anxious outrigger not yet hitched to the chariot team; however, a horse and rider are also plausible references, because both cavalry and chariotry were prevalent in ninth–sixth-century warfare.[12]

Despite the prominent imagery conjured by the Job 39 passage, it is the descriptions that are missing from the list of attributes of the excited war-horse that are most intriguing. It is strange that, although the ancient Hebrew poet discusses in detail the battlefield *sounds* and *smells* that excite the horse, he says nothing about what the horse *sees*. Remarkably, only a few verses after the quoted text (Job 39:19–25), in 39:29, the poet takes time to emphasize the superior eyesight of the eagle ("and her eyes behold afar off") and yet never mentions the horse's superior eyesight. In fact, the visual description of the horrific battlefield scenery and any effect it may have had on the war-horse is omitted. Did the author know that war-horses were often visually impaired by design and, thus, saw very little of the battle?

## The Sights of Battle

A horse is easily frightened by what it sees or imagines it sees; this is particularly true of the war-horse. Ancient battlefields reverberated with visual stimuli that

12. Drews, *Early Riders*, 48–59. For biblical references to mounted horsemen and chariotry in Iron Age warfare, see 1 Kgs 20:20–21; 2 Kgs 9:17–18, 21; Jer 46:4; Hos 14:3.

terrified horses, from waving banners, shining shields, and glittering spears to the elaborate headdresses on warriors and other horses.[13] Notably, the chariot races at the Olympic Games in Greece instituted in 680 included a specially devised "horse terrorizer" (*taraxippos*) located near the racetrack. The *taraxippos* was so frightening to the horses that they panicked and ran even faster, risking the wrecks that thrilled spectators.[14] Ancient battlefields were replete with naturally occurring "horse terrorizers," and the extraordinary sensory impact of those images had to be minimized before there could be any hope of controlling the horse in battle.

Understanding basic information about a horse's vision enhances our knowledge about the war-horse and, in particular, ancient chariotry. The horse's eye is twice the size of the human eye and larger than that of most mammals, including whales and elephants.[15] Unlike humans, who have binocular vision with a sight range of 180 degrees, the horse sees with both binocular and monocular vision which provides a total sight range of 345 to 355 degrees. Each eye has a separate field of view, and with each eye a horse can see to the front, to the side, and to the rear.[16] Despite enjoying such a wide range of vision, horses have four significant blind spots: directly behind, under the head, in front of the forehead, and on the back near the withers (shoulders).[17] Notwithstanding these inherent "blind spots," a horse with 20/30 vision has eyesight that is far superior to the average human with 20/20 vision.[18]

A horse's eye does not automatically focus like a human's eye. To a horse, landscapes tend to merge into one picture, similar to paintings designed to camouflage objects hidden in the background. Because horses have poor depth perception, separate objects only become distinct when they are in motion.[19] To focus on an object, a horse must move its head to change the position of its eye.[20] When its head is confined by a bridle and reins, a horse brings objects into focus by shying, leaping, dancing, and

13. The elaborate lion-head helmet as worn by Alexander the Great and his cavalrymen, as seen on the depiction of the Greek god Heracles on his fourth-century coinage, was designed to frighten the enemy horses when pulled over the face of the rider. The face of an approaching lion undoubtedly spooked the enemy horses and aroused their natural instinct to flee from predators (see fig. 2.1). The horse-hair plumes and elaborate regalia evident on the polls of the Bronze Age Egyptian, Neo-Hittite, and Iron Age Assyrian war-horses may also have been devised to frighten the enemy horses. For examples, see Y. Yadin, *The Art of Warfare in Biblical Lands* (trans. M. Pearlman; 2 vols.; New York: McGraw Hill, 1963) 1:216–17, 366–67, 386–87.

14. Although the exact description has not been found, the *taraxippos* was a standard feature of other hippodromes as well. See Pausanius 6.20.10–19 at http://www.theoi.com/Text/Pausanius6B.html. J. Swaddling, *The Ancient Olympic Games* (London: British Museum, 1999) 83–85.

15. C. Hill, *How to Think like a Horse: The Essential Handbook for Understanding Why Horses Do What They Do* (North Adams, MA: Storey, 2006) 18.

16. K. Wall, "Spooking and the Equine Eye," *Chronicle of the Horse*, 28 March 2003, 28–29.

17. Hill, *How to Think like a Horse*, 21–24.

18. E. Hanggi, "Seeing Eye to Eye," *Horse* 73, Summer 2003, 70–74.

19. Wall, "Spooking and the Equine Eye," 28.

20. A. Hogg, *The Horse Behaviour Handbook* (Newton Abbot, UK: David & Charles, 2003) 40.

"quivering like a locust," as described in Job 39:20. In battle, a sudden swivel of the body and movement of the head may bring a frightening object into focus.

To prevent their horses from seeing the chaos of the battlefield, ancient warriors sometimes adorned their eyes with "blinders," also known as "blinkers." Because the horse has a naturally occurring blind spot directly in front of its face, the addition of blinders limits a horse's field of vision to about 10–15 degrees per eye. Horses wearing blinders leap around less because they see fewer things to scare them. This partial blindness made ancient war-horses less likely to panic and more attentive to the direction of the charioteer.[21] Typically, blinders were constructed of hard leather or carved from ivory; more ornate blinders were bronze and covered in gold foil.[22] During excavations from 1949 to 1963 C.E., some 216 specimens of Assyrian horse blinders carved from elephant ivory were found at Fort Shalmaneser in Nimrud (Kalhu). The blinders varied from a flat *D*-shape to a longer spade-shape; they ranged in size from three to five inches in length to approximately three inches in width.[23]

The use of horse blinders in ancient warfare is well documented.[24] Horse blinders appear in Egyptian art as early as the Amarna period (ca. 1350), drawn on the chariot horses of Akhenaten (1358–1340), on a relief in the tomb of Pinhasy. Similar blinders are also illustrated on the chariot horses depicted in a battle scene on the Painted Box from Tutankhamun's tomb (ca. 1336). The Ostrich Hunt fan, also found in Tutankhamun's tomb, shows chariot horses adorned with gold-engraved blinders.[25] Elliptical horse blinders, artistically placed above the horse's eyes in a presumed attempt to emphasize facial expressions, appear on Abu Simbel reliefs depicting Ramesses II in the Battle of Qadesh (1285).[26]

Five hundred years later, carved ivories from the Fort Shalmaneser excavations in Nimrud show Assyrian chariot horses wearing blinders.[27] These ivory horse blinders,

21. The classic Hollywood scene of blindfolding the horses before leading them from the burning barn happens to be accurate and illustrates this principle.

22. J. H. Crouwel, "Chariots in Iron Age Cyprus," in *Selected Writings on Chariots* (Leiden: Brill, 2002) 162 (repr. of *Report of the Department of Antiquities Cyprus* [Nicosia: Dept. of Antiquities Cyprus, 1987]).

23. J. J. Orchard, *Equestrian Bridle-Harness Ornaments: Catalogue and Plates* (Ivories from Nimrud [1949–1963] 1/2; London: British School of Archaeology in Iraq, 1967) pls. 3, 22.

24. E. Gubel, "Phoenician and Aramean Bridle-Harness Decoration: Examples of Cultural Contact and Innovation in the Eastern Mediterranean," in *Crafts and Images in Contact: Studies on Eastern Mediterranean Art of the First Millennium B.C.E.* (ed. C. E. Suter and C. Uehlinger; OBO 210; Fribourg: Academic Press / Göttingen: Vandenhoeck & Ruprecht, 2005) 111–47. See also I. J. Winter, "North Syria as a Bronzeworking Centre in the Early First Millennium *B.C.*: Luxury Commodities at Home and Abroad," in *Bronzeworking Centres of Western Asia c. 1000–539 B.C.* (ed. J. Curtis; London: Kegan Paul, 1988) 193–225.

25. T. G. H. James, *Tutankhamun: The Eternal Splendor of the Boy Pharaoh* (New York: Metro, 2000) 36–37, 78–79, 186–87.

26. I. Brega, *Egypt, Past and Present* (New York: Barnes & Noble, 2001) 34–35.

27. D. Ussishkin dates the chariot ivories to the last quarter of the ninth century, about 800, based on the depiction of the chariot wheel with six rather than eight spokes. He concludes that the six-spoke

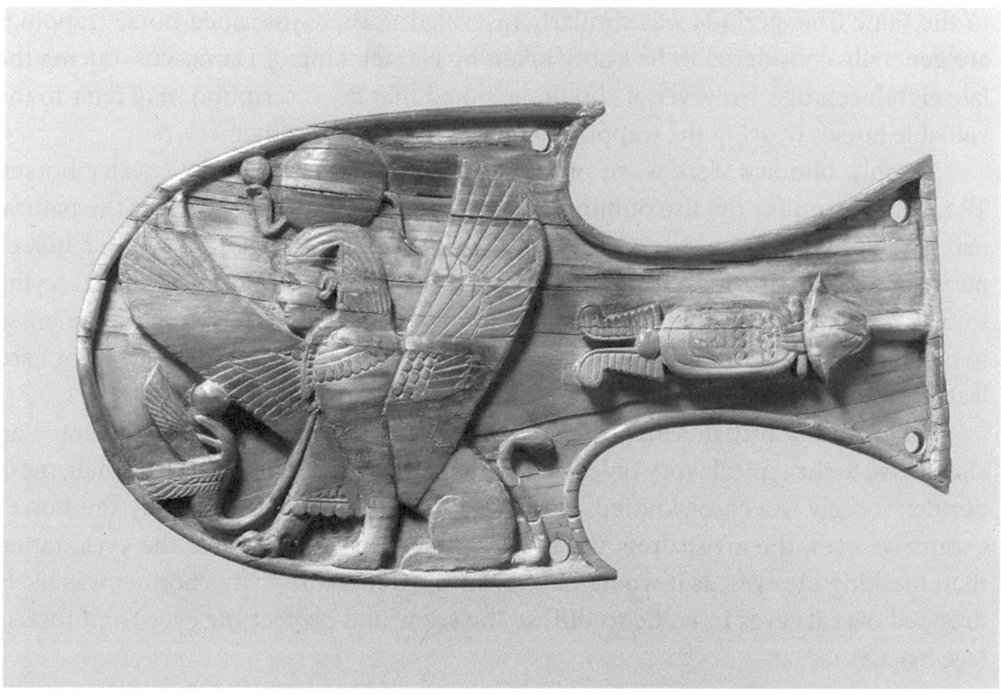

Fig. 2.2. Horse blinder with sphinx, 8th–7th century B.C.E.; Neo-Assyrian, Mesopotamia, Nimrud (ancient Kalhu). Ivory; H. 4.13 in. (10.49 cm). Rogers Fund, 1954 (54.117.1). Image © and published with the permission of the Metropolitan Museum of Art.

dating to the ninth and eighth centuries, are on display at the British Museum in London and the Bible Lands Museum in Jerusalem (see fig. 2.2). Several additional ivory and bronze eye blinders, also dating to Iron Age Israel, have been found at Tell el-Farʿah, Lachish, Beth Shemesh, and Megiddo.[28] They are shaped like flat soup spoons, and when attached to the horse's bridle block out all peripheral vision.

A particularly interesting late-eighth-century blinder inscribed in Aramaic "that which Hadad gave our lord Hazael from ʾUmqi in the year that our lord crossed the river" is on display in the National Archaeological Museum in Athens.[29] Additionally, a bronze frontlet from Samos depicting four figures of naked goddesses dating

wheel developed between the reigns of Shalmaneser III and Tiglath-pileser III ("On the Date of a Group of Ivories from Nimrud," *BASOR* 203 [1971] 22–27). Littauer and Crouwel disagree, dating the Nimrud chariot ivories to the late eighth century, about 720, during or after Sargon II, based on the four men in the chariot, which they view as a later development ("The Dating of a Chariot Ivory from Nimrud Considered Once Again," *BASOR* 209 [1973] 27–33).

28. MiYoung Im, "Archaeological Evidence of Horses and Chariotry in the Land of Israel during Iron Age II," *ASOR Newsletter* 56/1–2 (2006) 20.

29. I. Ephʿal and J. Naveh, "Hazael's Booty Inscriptions," *IEJ* 39/3–4 (1989) 192–200.

to the same time periods was similarly inscribed.[30] These inscribed horse trappings are generally considered to be booty taken by Hazael, king of Damascus, during the late eighth century. However, it should be noted that the inscription may refer to the valuable horses wearing the trappings and not the trappings themselves.

Notably, blinders were worn only by chariot horses and not by cavalry horses. The main reason for the use of blinders on chariot horses was to prevent the natural fear incited in horses by chariot wheels. Spinning chariot wheels, as seen in a horse's peripheral vision, created a frightening optical illusion of large whirling objects trying to overtake it. A horse's blinders blocked out the spinning wheels, forcing the horse to be more obedient to the driver and less distracted.[31] Blinders made of ivory or hard leather also protected the horse's eyes from enemy arrows and spears.

Reliefs in the British Museum suggest that the Assyrians may have invented an alternative to the typical ivory or leather horse blinder. The reliefs depict a small, mesh bonnet, loosely woven or knitted, sitting atop a horse's brow. To display the horse's expressive eyes, the artist drew the mesh bonnet pushed up above the eyes, rather than masking the eyes, as it would have been worn in battle.[32] This bonnet was likely dropped over its eyes in battle to diffuse the scene and protect the eyes from insects (see fig. 2.3).

The common use of blinders and mesh eye bonnets on chariot horses may be the simple answer to the question why the Job poet described only the smells and sounds affecting the war-horse but left unmentioned the sights of the battlefield that would have undoubtedly factored into the war-horse's performance. Perhaps it was common knowledge that war-horses went into battle wearing blinders that blocked nearly 90 percent of their vision.

## *The Noise of Battle*

Shouting men, neighing horses, blaring trumpets, anguished screams, clattering chariots, and ringing bells—the cacophony of battle noises—while merely loud to the human ear are heard in deafening, quadraphonic surround sound by the horse. In addition to their superior sight, horses have extremely sensitive hearing. A horse has 16 muscles controlling each of its funnel-shaped ears, usage of which makes sounds

30. See D. Parayre, "A Propos d'une plaque de harnais en bronze découverte à Samos: Réflexions sur le disque solaire ailé," *RA* 83 (1989) 45–51; see also Crouwel, "Chariots in Iron Age Cyprus," 162–64.

31. Carriage horses today still wear blinders to block out frightening traffic sights. Racehorses ridden by jockeys typically do not wear blinders because their trainers want them to know exactly where the horses behind are and how fast they are gaining on them. In harness racing, however, blinders are optional but used by most trainers because the wheels of the pursuing sulkies are distracting to the horses. T. Grandin and C. Johnson, *Animals in Translation* (Orlando: Harcourt, 2005) 40–41.

32. A. H. Layard, *The Monuments of Nineveh* (London: John Murray, 1853; repr. Piscataway, NJ: Gorgias, 2004) pl. 23.

Fig. 2.3. Eye bonnet protectors. Stone panel from the North Palace of Ashurbanipal. British Museum 0017534001. © The Trustees of the British Museum. Published with permission.

seem clearer and louder.[33] By moving each ear separately toward a sound, horses can locate the sound's *exact* source.

Because a horse can move its ears in every direction around its body, it can hear approaching danger much sooner than a human and before either can see the origin of the sound. When strange sounds startle a horse, it will prick its ears toward the sound and prepare for flight, or in the case of the seasoned and brave war-horse, prepare to run toward it. Running toward battle instead of fleeing is an anomaly and is something the horse must be trained to do. Some ancient horses undoubtedly failed to overcome their natural instincts and fears and never made suitable war-horses.[34]

33. J. Kidd, ed., *The Way of the Horse: How to See the World through His Eyes* (New York: Howell, 1998) 44.

34. "But one should beware of horses that are naturally shy. For timid horses give one no chance of using them to harm the enemy, and often throw their rider and put him in a very awkward situation" (Xenophon, *Art of Horsemanship* 3.9).

Ancient metaphors for the sounds of an approaching army were based on the loudest noises known from nature: the roaring sea (Jer 6:23, 50:42; Isa 5:30), a roaring lion (Isa 5:29), and a whirlwind (Jer 4:13, Isa 66:15). More naturalistically, they also included noises associated with chariot warfare: "Why are his chariots so long in coming? Why so late the *clatter of his wheels*?" (Judg 5:28; emphasis added); "The crack of whips, *the clatter of wheels*, galloping horses and jolting chariots!" (Nah 3:2 [NIV]; emphasis added); "With a *clatter as of chariots* they bound on the hilltops, with a *noise like a blazing fire consuming straw*; like an enormous horde arrayed for battle" (Joel 2:5; emphasis added); "from the *clatter of horsemen and wheels and chariots, your walls shall shake*" (Ezek 26:10; emphasis added); and "Men shall cry out, all the inhabitants of the land shall howl, at *the clatter of the stamping hoofs of his stallions, at the noise of his chariots, the rumbling of their wheels*" (Jer 47:2–3; emphasis added).

A number of the noises associated with ancient battles, such as screams of pain from the wounded and dying, the neighing of frightened horses, and the clattering of the chariots, were the involuntary yet predictable aspects of deadly conflict.[35] Some sounds, such as trumpet blasts and ringing bells served specific purposes of organizing people and horses for battle. Still other sounds, such as the noise made by clanging spears against shields were made by the enemy in calculated attempts to frighten the opposing horses, having desensitized their own horses to the particular sound.[36]

Part of any war-horse's training included systemically desensitizing it to noise.[37] Precombat training was absolutely essential. In modern times, mounted police subject their horses-in-training to "sound sessions" with sirens, crowd noise, horns, truck engines, and gun shots to prepare them for law enforcement duties. In the ancient world, desensitization to shouts and trumpet noise was particularly important, because both were constantly present on the battlefield. Xenophon suggested that grooms lead young colts through crowds as a means of accustoming them to an ever-changing array of sights and noises.[38] A specific part of the training suggested by Xenophon was: "If a shout is heard or a trumpet sounds, you must not allow the horse to notice any sign of alarm in you . . . but as far as possible let him rest in such circumstances, and, if you have the opportunity, bring him his morning or evening meal."[39]

35. "The snorting of his horses was heard from Dan: the whole land trembled at the sound of the neighing of his strong ones; for they are come, and have devoured the land, and all that is in it; the city, and those that dwell therein" (Jer 8:16, KJV).

36. Xenophon mentions this battle strategy used by the Greeks against the Persians in the Battle of Kounaxa (400) shortly before Cyrus was killed (*Anabasis* [trans. C. L. Brownson; LCL 90; Cambridge: Howard University Press, 1998] 1.8).

37. Some riding horses can never become acceptable harness horses because they cannot tolerate noises under harness. It is vital for everyone's safety that horses are exposed to a vast array of noises during training. H. Bean and S. Blanchard, *Carriage Driving: A Logical Approach through Dressage Training* (New York: Howell, 1992) 115–18. Even so, some drivers resort to stuffing cotton in their horses' ears as a precaution (Hill, "How to Think," 28).

38. Xenophon, *Art of Horsemanship* 2.5.

39. Ibid., 9.11.

Fig. 2.4. Bell inscribed with the Urartian royal name Argishti, 786–756 B.C.E. Eastern Anatolia or northwestern Iran, Urartian bronze, iron; 3.43 in. (8.71 cm). Gift Nathanial Spear, Jr., 1977 (1977.186). Published with the permission of the Metropolitan Museum of Art.

In the ancient world, part of the war-horses' battle regalia included bells attached to their harnesses. Horse bells previously attached to harnesses were excavated in situ with chariot horse skeletons found at the Hasanlu stable site (ca. 800 B.C.E.) in Iran.[40] The biblical text also notes the prevalence of horse bells during the late Iron Age in Zech 14:20, "In that day, even the bells on the horses shall be inscribed 'Holy to the Lord.'" A horse bell from Urartu, dating to 786–756, and now on display in the Metropolitan Museum in New York bears the cuneiform inscription "from the arsenal of Argishti." An inscription of this sort suggests that the bells were used as identifiers, resembling modern-day dog tags, to facilitate the return of a horse to its proper location[41] (see fig. 2.4).

Conceivably, the bells, made of bronze or iron, were also affixed to the inside of the tassels on the horse's regalia, where they would jingle to assist in warding off

40. Apparently, the stables at Hasanlu burned quickly; at least nine horses were not evacuated. Their skeletons indicate that the horses were small, with an average withers height of about 13.5 hands. Horse gear found in the stables and tack room includes headstalls, bits, harnesses, decorated breastplates, chamfrons, and bronze bells. M. de Schauensee, "Horse Gear from Hasanlu," *Expedition* 31/2–3 (1989) 37–52.

41. See "Bell Inscribed with the Urartian Royal Name Argishti [Eastern Anatolia or Northwestern Iran, Urartian] (1977.186)," *Timeline of Art History* (New York: Metropolitan Museum of Art, 2000), http://www.metmuseum.org/toah/ho/04/waa/ho_1977.186.htm (accessed October 2006).

biting insects. Perhaps the bells jingling in unison from each horse also functioned somewhat like a metronome to govern the proper cadence in a particular gait.[42] When two or three chariot horses were yoked together, it was important that their stride be in synch, allowing the chariot to move smoothly over the ground. Additionally, the bells, although not loud to the human ear, probably created a white noise, a reassuring and familiar sound on the battlefield that lessened the heinous sounds of death for the horse. Many sleigh horses still wear bells today for these precise reasons. The jingle bells on the harnesses are also used to announce a horse's arrival on roadways and to alert others to the presence of horse-drawn vehicles. The bells were required by law in many jurisdictions, just as horns are now required on cars.[43]

## *The Smells of Battle*

"From afar he smells the battle" (Job 39:25). The horse's sense of smell is far more acute than a human's. It acts as an early-warning system for danger.[44] When alarmed, a horse flares its nostrils, sniffing toward the perceived threat.[45] Horses use their sense of smell to investigate strange objects and as a recognition tool to remember other animals and people. With a light breeze, a horse can smell a single mare in season from over a half mile away.[46] Likewise, a horse can sense the unfamiliar smell of an opposing army's horses, mules, asses, and camels from over a mile away, long before they can be seen.[47] The horse's physical reaction to such unpleasant and unfamiliar

42. Horses love music, and some will dance to the beat naturally, while others may be trained to do so. E. LeGuin, "Man and Horse in Harmony," in *The Culture of the Horse: Status, Discipline, and Identity in the Early Modern World* (ed. K. Raber and T. J. Tucker; New York: Palgrave Macmillan, 2005) 193. The musical freestyle, a dance to a music routine by mounted riders, is now part of the Olympic competition in dressage. The second-century C.E. historian Aelian tells the story of horses in Sybaris, Italy, who were trained to dance in time to pipe music for banquets. The crafty enemy learned this fact and, when they were within bowshot, had players strike up dance music, whereupon the horses shook off their riders and began to leap about and dance. Aelian comments that "they not only threw the ranks in confusion but also 'danced' away the war" (Aelian, *On Animals* 16.23 [295]).

43. M. Hull, *The Horse in Harness* (Philadelphia: Chelsea, 2002) 38.

44. Kidd, *The Way of the Horse*, 52.

45. Hogg, *Horse Behaviour*, 48.

46. D. Morris, *Horsewatching* (New York: Crown, 1988) 36.

47. Camels have a particularly bad odor for horses unaccustomed to them. "The horse fears the camel and cannot abide the sight or smell of it," Herodotus notes in describing the successful battle strategy of Cyrus in spooking the Lydian cavalry horses with a camel charge by his former pack animals at the Battle of Sardis. "Indeed, as soon as the battle was joined, the very moment the horses smelled the camels and saw them, they bolted back; and down went all the hopes for Croesus" (*The Histories* [trans. R. Waterfield; Oxford: Oxford University Press, 1998] 1.80). In fact, camels were typically kept in the back of army processions to avoid irritating the horses (Herodotus, *Histories* 7.87).

The U.S. Army, after an eight-year experiment (1855–1863 C.E.), gave up the idea of a camel brigade because the horses reacted so badly to the camels. R. Froman, "The Red Ghost," *American Heritage* 12/3 (1961) 35–37, 94–98. War camels ridden by Arab soldiers are prominent in the Assyrian battle scenes on the reliefs from Nineveh (ca. 645). J. E. Reade, *Assyrian Sculpture* (Cambridge: Harvard University Press,

smells, usually a snort, a stomp, or a toss of the head, can serve to alert its rider to an approaching threat.[48]

Once a battle is underway, the additional odors of blood, urine, dung, and the elusively charged smell of fear from both humans and beasts add to the startling scents detected by the anxious war-horse. The biblical text also mentions the scent of fire from burning chariots and the shrouds of dust rising from battlefields: "He maketh wars to cease unto the end of the earth; he breaketh the bow, and cutteth the spear in sunder; he burneth the chariot in the fire" (Ps 46:9, KJV); "I will burn down her chariots in smoke" (Nah 2:14); "from the cloud raised by his horses dust shall cover you" (Ezek 26:10).

According to the tomb biography of Amenemheb, an Eighteenth-Dynasty Egyptian army officer, the scent of a mare in season distracted the Egyptian stallions in a fifteenth-century chariot battle pitting the Egyptians, led by Thutmose III (1479–1425), against the Canaanites at Megiddo. The heroic effort of Amenemheb in killing the offending mare saved the day by allowing the stallions to return to their business of war without further distraction.[49] At first glance, this story may seem apocryphal, because a mare in season would have disturbed stallions of either forces, and killing her would not necessarily have eliminated the scent. However, if as claimed by Amenemheb, the mare was released as a battle tactic to disrupt the Egyptian stallions, it is likely that the nostrils of the Canaanite stallions were swabbed with a potent-smelling salve to mask the smell of the mare.[50] Whether Amenemheb tried to eliminate the scent by burning or burying the mare remains unknown, but his story reveals a logical battle tactic and a likely dilemma on any ancient battlefield.

It is also true that horses can smell fear, especially that of other horses. This emotion is particularly interesting in the context of war, where the "fear factor" is greatly elevated. Scientific evidence suggests that, in addition to possessing certain innate fears, horses learn and imitate fear from other horses.[51] Horses secrete acrid sweat and

---

1999) 88, pl. 105. Camels ridden by the Arab contingent fought in the Battle of Qarqar (853). S. Yamada, *The Construction of the Assyrian Empire: A Historical Study of the Inscriptions of Shalmaneser III (859–824 B.C.) Relating to His Campaigns to the West* (Culture and History of the Ancient Near East 3; Boston: Brill, 2000) 157. In the biblical text, camels are mentioned in a military context in connection with the invading Midianites (Judg 6:1–6). We may assume that most Iron Age war-horses in Israel and Assyria were desensitized to camels.

48. In male horses, the scent of a mare in season usually generates the "flehmen" response, a peculiar facial gesture wherein the horse raises its head and curls its upper lip backwards toward its nostrils to trap the fragrance. Hogg, *Horse Behaviour,* 48. See also Jer 5:8.

49. A. R. Schulman, "Military Organization in Pharaonic Egypt," in *CANE* (ed. Jack M. Sasson; 4 vols.; New York: Scribner's, 1995) 1:289–302.

50. The practice of rubbing the nostrils of stallions with Vicks VapoRub® or other strong-smelling aromatic oils to help them behave remains common today (Morris, *Horsewatching,* 37).

51. Also, there is evidence that the more intelligent breed of horses are the most curious and most fearful (Grandin and Johnson, *Animals in Translation,* 40–41).

droppings when frightened. The pungent odor of these substances has the biological effect of making other horses wary and unsettled.[52] There is no doubt that the multifarious scents of battle keenly affected horses and added to their nervousness.

### *The Sex of the War-Horse*

With regard to the preferred sex of the war-horse in antiquity, it is clear that stallions were admired for their qualities of strength and virility. Assyrian reliefs clearly show stallions pulling the battle chariots.[53] However, the consistent depiction of stallions in Assyrian art may have been more of an artistic convention, given that most of the placement of horse genitalia is anatomically incorrect (that is, located farther toward the center of the horse, and thereby more visible than they appear in reality; see fig. 2.5).[54] It is just as likely that ancient chariot horses were geldings, as were the fifth-century chariot horses buried in Pazyryk.[55]

It is also certain that mares were sometimes used in chariotry, as is evident from fourteenth-century Nuzi texts and the later biblical text: "I have likened you, my darling, to a mare in Pharaoh's chariots" (Song 1:9, LXX).[56] Pregnant mares are sexually neutral and are not distracting to stallions. It is possible that after the foals were weaned, usually six months after birth, mares could be used in the chariotry until they approached their next birthing date about five months later. However, most broodmares are designated for breeding only and not are used as performance horses. The practice in ancient times probably varied depending on the availability of trained chariot horses and the numbers needed for each campaign.

The numerous letters and administrative reports from Assyria during the Iron Age that differentiate between breeding stallions (*urū*), riding horses (*ša pēthalli*), broodmares (*ú-rat*.MEŠ), and riding mares (*ú-rat pít-ḫal-lum*) support this observation.[57] Riding horses far outnumbered those designated as stallions. Thutmose III's booty list from the Battle of Megiddo (1479) also states that he took "2,041 horses, 191

52. Hogg, *Horse Behaviour,* 48.

53. Reade, *Assyrian Sculpture*, 74.

54. For the suggestion that stallions depicted frequently in Greek art may be an artistic convention designed to heighten the power and majesty of the rider, see I. G. Spence, *The Cavalry of Classical Greece: A Social and Military History with Particular Reference to Athens* (Oxford: Clarendon, 1993) 44.

55. J. Clutton-Brock, *Horse Power: A History of the Horse and the Donkey in Human Societies* (Cambridge: Harvard University Press, 1992) 100.

56. A. Azzaroli, *An Early History of Horsemanship* (Leiden: Brill/Backhuys, 1985) 29. The first-century C.E. historian, Aelian, also records that "mares are believed to be most suitable for drawing chariots" (*On the Characteristics of Animals* 11.407).

57. For translations of numerous tablets from Sargon's reign that record the receipt of horses from various countries, including Syria and Israel, see B. Parker, "Administrative Tablets from the North-West Palace, Nimrud," *Iraq* 23 (1961) 26, 46 (n. 3), 58, and 60. For further comment on the kinds of horse-list texts from Nimrud, see "An Assyrian Horse List," trans. K. L. Younger Jr. (*COS* 3:128: 279–80) (this text is also known as *TFS* 99).

Fig. 2.5. Inside the Assyrian camp Nimrud (ancient Kalhu), northern Iraq. Neo-Assyrian, 883–859 B.C.E. Stone panel from the Northwest Palace of Ashurnasirpal II (Room B, Panel 7) 00032393001. © The Trustees of the British Museum. Published with permission.

foals, 6 stallions, and . . . colts," indicating a differentiation between breeding stallions and regular chariot horses.[58]

Notably, in the biblical text, the word translated 'stallion' is the Hebrew word אברים, meaning 'mighty ones'. This term is variously translated 'stallions' or 'steeds' and appears only in Judg 5:22 and in the book of Jeremiah (8:16, 47:3, 50:11). In these contexts, the term always refers to loud noises made by war-horses, such as pounding hooves or neighing—which are, of course, not gender specific.[59] 'Mighty ones' could

58. "From The Annals of Thutmose III" (Lichtheim, *AEL*, 2:33). See also R. Drews, *The End of the Bronze Age: Changes in Warfare and the Catastrophe ca. 1200 B.C.* (Princeton: Princeton University Press, 1993) 130.

59. I am aware that the etymology of the word often denotes the male gender because the Egyptian word for 'stallion' *ʾibr* is possibly a loanword from Canaanite (*ʾibr* in Ugaritic), and its derivatives include אבר 'member, penis'. However, because male is the default gender in most

be a reference to trained war-horses in general, not particularly stallions, because the sex of the horse is generally unrelated to the amount of noise it makes. Therefore the translation 'mighty war-horses' may be more appropriate than 'stallions'.

In all likelihood, a large number of the horses used for chariotry and riding were geldings, not stallions. Geldings, castrated male horses, possess a calmer temperament, which allows trainers to work them together and stable them side-by-side with mares and other geldings. This suggestion, of course, presumes that castration was practiced as routinely on horses as it was on cattle (oxen) and humans (eunuchs).[60] The earliest textual reference to castration of horses appears in an Old Hittite legal text dating to the sixteenth century and seems to indicate the superior value of a breeding stallion to gelded horses:

> If anyone finds a stallion and castrates it, when its owner claims it, the finder shall give 7 horses: 2 two-year-olds, 3 yearlings, and 2 weanlings. (Old Hittite Law 61)[61]

This Hittite law suggests that most horses were routinely gelded when young, and only a few superior colts were left complete as breeding stallions. The historian Strabo (63–24 C.E.) provided another textual reference to castrated horses, when he mentioned that the sixth-century B.C.E. Scythians castrated their horses to make them easier to manage.[62] Xenophon, writing in the fourth century B.C.E., confirms that "vicious horses, when gelded, stop biting and prancing around, to be sure, but are none the less fit for service in war."[63] Nevertheless, Greek art typically portrays stallions as war-horses, and they may well have been preferred by cavalrymen.[64]

---

languages, and these contexts refer to noises that both male and female horses as well as geldings can make, I prefer a more gender-neutral translation. See C. H. Gordon, *Ugaritic Textbook: Grammar, Texts in Transliteration, Cuneiform Selections, Glossary, Indices* (rev. ed.; AnOr 38; Rome: Pontifical Biblical Institute, 1998) no. 39; see also W. F. Albright, *The Vocalization of the Egyptian Syllabic Orthography* (New Haven, CT: American Oriental Society, 1934) 33, 1; and אביר in *HALOT*.

60. Some argue that horses were too valuable to risk castration in antiquity because the lack of medical sophistication and hot climate would result in a high mortality rate. See A. Hyland, *Equus: The Horse in the Roman World* (New Haven, CT: Yale University Press, 1990) 81. However, castration of horses carried no greater risk of infection than castrating cattle or humans, both of which were castrated routinely. In fact, the recovery period from castration surgery on horses is surprisingly brief—7 to 10 days. The younger the colt is when castrated, the less traumatic the process and the quicker the recovery. E. L. Squires, *Understanding the Stallion* (Lexington, KY: Eclipse, 1999) 46–47. Today some breeders prefer to castrate their colts as weanlings (six months); others wait until they are yearlings. The older a horse is when castrated, the more certain he is to retain "studsy" characteristics that are problematic around other horses.

61. M. T. Roth, *Law Collections from Mesopotamia and Asia Minor* (2nd ed.; SBLWAW 6; Atlanta: Scholars Press, 1997) 226.

62. Strabo, *Geography* (trans. H. L. Jones; LCL; New York: Putnam, 1917–32) 7.4.8.

63. Xenophon, *Cyropaedia* (trans. W. Miller; LCL 52; Cambridge: Harvard University Press, 2000) 7.5.62.

64. Spence, *Cavalry of Classical Greece*, 44.

## The Horse in Battle

The role of the ancient war-horses was pivotal in the conduct of battle.[65] Once a battle began, chaos reigned. Ancient trainers strained to gain enough control of their horses for them to be useful on the battlefield and functional as weapons. Even so, the problem of the panicked horse forever loomed. The *Iliad* describes seven chariot crashes, five of which were caused by panicked horses.[66] In fact, ancient battle strategies often included plans to try to panic the enemies' horses.[67] The panic and chaos of battle is recurrently depicted on Egyptian and Assyrian reliefs; it is also chronicled in literature.[68] For example, after the battle with the Elamites in 691, Sennacherib boasts of his battle prowess and the bravery of his horses:

> I c[ut open their necks like sheep, I slit th]eir precious [thr]oats [like thread. Like a mighty flood of the storm] season [. . .] I caused [the]ir blood to flow down [upon the broad earth. My swift] thoroughb[reds], the team of my cha]riot, [plunged] into [the mass] of their blood [as (though into) a river (and) the wheels of my battle-chariot] [(which) overthrows rogues and criminals] were bathed in blood and filth.[69]

Sennacherib's description of horses and chariots running wildly on the battlefield is similar to a seventh-century account attributed to the prophet Nahum in the biblical text:

> The chariots storm through the streets,
> Rushing back and forth through the squares.

65. For a detailed discussion of the role of the war-horse in specific battles from Classical Greece to medieval knights, see Philip Sidnell, *Warhorse: Cavalry in Ancient Warfare* (London: Hambledon Continuum, 2006). See also Louis A. DiMarco, *War Horse: A History of the Military Horse and Rider* (Yardley, PA: Westholme, 2007), which includes a chapter on the World War I battles fought in southern Israel and the Megiddo Plain (pp. 309–49). For a fascinating account of the recent use of horses in mounted combat against the Taliban and El Qaeda by U. S. Special Forces and Afghanistan forces, see Doug Stanton, *Horse Soldiers: The Extraordinary Story of a Band of U. S. Soldiers Who Rode to Victory in Afghanistan* (New York: Scribner, 2009).

66. For example, "Rearing, bolting in terror down the plain his horses snared themselves in tamarisk branches, splintered his curved chariot just at the pole's tip and breaking free they made a dash for the city walls where battle-teams by the drove stampeded back in panic," *The Iliad* 6.45–50. See also *Iliad* 8.350–60, 11.149–88, 15.410–20, 16.435–50.

67. See Zech 12:4, "'On that day I will strike every horse with panic and its rider with madness,' declares the Lord." For the interesting account of men panicking and leaving their horses in haste at the supposed sound of the chariots of an invading army, see 2 Kgs 7:6.

68. Although the accuracy of chaotic battle descriptions is sometimes questioned, I find it interesting that the Assyrians had official court reporters who accompanied the king to battle. It is reasonable to assume that the Israelites did as well. Certainly there were enough surviving participants to know if battle descriptions proclaimed by the biblical prophets rang true. On balance, I believe that we may trust these battle descriptions of chaos and destruction as fairly accurate.

69. A. K. Grayson, "The Walters Art Gallery Sennacherib Inscription," *AfO* 20 (1963) 83–96, lines 88–91.

They look like flaming torches;
They dart about like lightning.

The crack of whips, the clatter of wheels,
Galloping horses and jolting chariots!

Charging cavalry,
Flashing swords and glittering spears!
Many casualties, piles of dead,
Bodies without number,
People stumbling over the corpses. (Nah 2:5, 3:2–3, NIV)

Although the Israelites left no pictorial evidence of warfare (as discovered to date), both Egyptian and Assyrian artwork consistently depicts chaotic battles, most of which include images of warriors being trampled by horses, as well as occasional drawings of wounded horses and wrecked chariots.[70] Both Assyrian and Aramean inscriptions and monuments recorded battles in which the horses fielded by the Israelite army numbered in the thousands, as discussed in chap. 3, below.

War-horses in the ancient world were more than battlefield transports or fast "get-away" devices; many were lethal weapons.[71] People feared horses as killing machines. Notwithstanding Greek myths about flesh-eating mares, literary and artistic evidence from the ancient world strongly suggests that war-horses were trained and/or forced to kill.[72] For example, corpses appear under the chariot horses (or onagers) in the Standard of Ur war panel, in the reliefs on the sides of the Eighteenth-Dynasty chariot found in the tomb of Thutmose IV (1411–1397), in the ivory plaque of chariots from Megiddo (thirteenth century), under the Neo-Hittite chariot horse in the orthostat from Carchemish (ninth century), and, most frequently, on the Neo-Assyrian reliefs with war scenes from Ashurnasirpal (ninth century) to Ashurbanipal (seventh

70. See Yadin, *Art of Warfare*, 1:192–93, 216–17; 2:386–87; Reade, *Assyrian Sculpture*, pls. 96, 82.

71. Whether chariots were used to shock the enemy by charging or merely as mobile firing platforms has sparked a good deal of academic debate yet remains unresolved. There are interesting arguments on both sides, as discussed more fully in chap. 4, below. For a review of the literature, see Drews, *The End of the Bronze Age*, 126–29; and for "shock cavalry," see Drews, *Early Riders*, 58–59, 71. Drews supports the argument that the chariots were primarily mobile platforms for archers but notes, "The whole point of the battle (as Egyptian reliefs show clearly enough) was to bring down as many of the opponent's chariots as possible." I agree. However, having seen horses push and trample people in crowded parades and having witnessed several modern-day carriage runaways and the consequential carnage, I believe excited battle horses overtook and ran over the enemy as a routine part of warfare. In fact, Xenophon's description of the Persian scythe-bearing chariotry mentions that the scythes "were also set *under* the chariot bodies, pointing towards the ground, so as to cut to pieces whatever they met." He explains further that the intention was for them to drive into the ranks of the Greeks as they advanced (*Anabasis* 1.8.10). See also *Cyropaedia* 7.1.28, 32, 47.

72. According to Greek legend, the mares of Diomedes ate human flesh but after devouring their dead master, became tame; the mythical Glaucus was torn to pieces by his mares. In both myths, the mares were made mad by the springs from which they drank (Aelian, *On Animals* 15.25).

century).[73] Wounded enemies also appear under the hooves of horses on the Balawat Gate depicting Shalmaneser's army (ninth century), and in a similar position under the mounted warriors and the chariot horses in the war reliefs from Sargon's palace at Khorsabad (eighth century).[74] This artistic motif used in battle depictions from the Bronze Age to the Iron Age clearly emphasizes the power of the chariot horse to knock down and trample the enemy.

Was this motif merely an artistic convention? I think not. Although there is room for disagreement on this topic, it is hardly an exaggeration to speculate that as many warriors were killed by the crushing blows of horses' hooves landing on vital body parts as by the random wounds of arrows and spears piercing the flesh. Typically, arrows were not lethal against armored soldiers unless they managed to hit an artery or a seam in the armor.[75] Even then, the wounded either bled to death like Ahab (1 Kgs 22:35) or later died of infection, as Josiah presumably did (2 Chr 35:23–24).

Surprisingly, the Greek mercenary and cavalry officer Xenophon is sometimes cited for the notion that horses never killed anyone in battle. This view is based on a statement he made that is generally quoted out of context. In an attempt to rally the defeated remnant of Cyrus's Greek army that was trapped in Persia without its horses (and outnumbered by the enemy's 10,000 horse cavalry), Xenophon proclaimed: "for nobody ever lost his life in battle from the bite or kick of a horse, but it is the men who do whatever is done in battles" (*Anabasis* 3.2). These words of encouragement served as an apparent attempt to ease the fear and dismay of the recently defeated (and now horseless) Greek troops who were understandably terrified of the Persian cavalry. Xenophon, an accomplished equestrian and outstanding military leader, further exhorted his troops by suggesting that the Persian cavalrymen were seated precariously on their horses and afraid of falling off. This battle rhetoric is wholly at odds with his writings after retirement, where he extols the virtues of Persian horsemanship and admits that superior horsemanship is the only way to win against a stronger army.[76]

The proposition that horses inflicted death blows in battle is also supported by ancient literary evidence. For example, the battle inscription of Ramesses III (1187–1156), which describes the war against the Sea Peoples (ca. 1187), reads: "The horses were quivering in every part of their bodies, prepared to *crush the foreign countries under their hoofs*."[77] According to the Deuteronomistic History, Queen Jezebel of Israel dies

73. Yadin, *Art of Warfare*, 1:132–33, 192–93, 242–43; 2:366, 382–85, 442–43.

74. Ibid., 2:396–97, 416–17.

75. R. A. Gabriel and K. S. Metz, *From Sumer to Rome: The Military Capabilities of Ancient Armies* (Contributions in Military Studies 108; Westport, CT: Greenwood, 1991) 93.

76. Xenophon, *On the Cavalry Commander* (trans. E. C. Marchant and G. W. Bowersock; LCL; Cambridge: Harvard University Press, 2000) 8.1; idem, *Art of Horsemanship* 6.12, 8.6. One victim, caught under the hooves of Cyrus's horse, managed to strike the horse in the belly with his sword (idem, *Cyropaedia* 7.1.37).

77. J. A. Wilson, trans., "The War against the Peoples of the Sea," *ANET*, 262.

Fig. 2.6. Greek Coin from Paeonia. Mounted warrior spearing a fallen enemy. Tetradrachm, 340–315 B.C.E. Author's collection.

in Jezreel under the hooves of Jehu's chariot horses: "They threw her down; and her blood spattered on the wall and on the horses, and *they trampled her*" (2 Kgs 9:33; emphasis added). The biblical text also identifies horses as tools of God's punishment with *hooves as sharp as flint* (Isa 5:28) and mentions death by trampling underfoot: "you were left lying unburied, like a *trampled corpse* [in] the clothing of slain, gashed by the sword" (Isa 14:19; emphasis added). Presumably from the late eighth century or early seventh century, the prophet Micah writes, "For I will give you horns of iron and provide you with *hooves of bronze and you will crush the many peoples*" (Mic 4:13; emphasis added).[78]

78. This interesting reference to "hooves of bronze" does not suggest that horseshoes were known as early as the eighth century. Xenophon's advice (ca. 400) on toughening hooves by standing on rough stones suggests that unshod horses were the norm (*Art of Horsemanship* 4.3). Archaeological evidence also suggests that metal horseshoes were not invented until the second century and did not come into common usage until the fifth century C.E. See G. Ward, "The Iron Age Horseshoe and Its Derivatives," *Antiquaries Journal* 21 (1941) 7–27; J. B. Ward-Perkins, "The Iron Age Horseshoe," *Antiquaries Journal* 21 (1941) 144–49; M. Littauer, "Early Horseshoe Problems Again," *Antiquity* 42/167 (1968) 221–25.

Horses' hooves grow up to a quarter of an inch per month. Modern-day domestic horses undergo hoof trimming and shoeing every six weeks. However, horseshoes are not necessary for sound feet. In the wild, their hooves wear off on the abrasive terrain as fast as they grow (Hill, *How to Think like a Horse,* 54). Many horses today—for example, the famous Lipizzan stallions at the Spanish Riding School in Vienna—are ridden unshod. Some horsemen contend that metal shoes keep the hoof from functioning normally by interfering with proper flexion, hindering circulation, and unbalancing the foot. For more information about the benefits of the "natural hoof," see Heather Smith Thomas, "What Can We Learn from the Natural Hoof?" *The Chronicle of the Horse*, 3 March 2000, 41–42.

Generally dated to the late seventh century, the prophet Habakkuk describes the Babylonian army horses as "swifter than leopards, *fiercer than wolves at dusk*" (Hab 1:8, NIV; emphasis added); Isaiah describes the Assyrian army as making Israel like "a thing *trampled* in the mire of the streets" (Isa 10:6; emphasis added); and Ezekiel mentions the horses in the army of Nebuchadrezzar as *trampling* the streets of Tyre (Ezek 26:11). The story of a horse fighting a man is found in 2 Macc 3:25 (emphasis added): "Charging furiously, the *horse attacked Heliodorus with its front hooves.*"

Greek coinage from fourth-century Thessaly portrays a rider spearing an enemy to the ground with a lance while the horse tramples his body.[79] A similar motif of a rider trampling a fallen enemy appears in coinage from Paeonia, struck for the king of Paeonia (335–315 B.C.E.; see fig. 2.6). According to the Roman historian, Aelian (170–230 C.E.), the Persian army specifically trained its horses to trample the enemy: "And they throw dummy corpses stuffed with straw beneath their feet in order that they may get used to trampling on corpses in war and may not through terror at some unnerving occurrence be useless in encountering men-at-arms."[80] Regardless of Aelian's reputation for writing more for the audience's entertainment than historical accuracy, abundant evidence from both art and literature supports the notion that horses indeed trampled people during battle.

Despite the disagreement over whether people were actually killed by horses during war maneuvers, the fear of being trampled to death was real enough in the ancient world that it developed into a literary motif designed to evoke fright and dread.[81] What is certain is that horses had to be trained to wound and kill humans because, as a prey animal, their first instinct is to flee, not to fight.[82]

## *The Limitations of the Horse in Battle*

Horses, unlike oxen, will not quit when they are tired.[83] They continue working until they die or collapse in exhaustion. It is the responsibility of a horse's trainers and

79. D. R. Sear, *Greek Coins and Their Values*, vol. 1: *Europe* (London: Seaby, 1978) 207, pl. 2166.

80. Aelian, *On Animals* 16.25.

81. Today, rodeo bronco riding and bull riding are extremely dangerous sports because of the likelihood that an animal might accidentally trample a fallen rider. It is easy to imagine the damage that a horse trained to trample could inflict. Ancient battle armor might deflect arrows and spears but would be almost useless against the 1,100-lb. (498.95-kg) weight of a stomping horse.

82. Today in the Olympic sport of dressage, gymnastic movements on horseback are based on ancient battle maneuvers. The highest levels require execution of the *piaffe*, a jogging/stomping in place ultimately designed to trample the enemy. For a description of specific battle maneuvers taught horses in the military academy in France, first opened in 1594 C.E., which included rearing, pawing, lunging, stamping in place, and kicking out behind, see K. Van Orden, "From *Gens d'armes* to *Gentilshommes:* Dressage, Civilité, and *Ballet à Cheval*," in *The Culture of the Horse: Status, Discipline, and Identity in the Early Modern World* (ed. K. Raber and T. J. Tucker; New York: Palgrave MacMillan, 2005) 202.

83. "One would think a tired horse would begin to quit . . . not necessarily . . . sometimes a very fatigued horse won't quit because he is just not thinking . . . and that can be extremely dangerous"; see

riders to conserve its energy. In modern equestrian sports, riders change polo horses every seven and one-half minutes; Olympic show-jumping and dressage competitions last only three to four minutes; and horse races are completed in less than two minutes. Even the Pony Express riders (1860–61 C.E.) changed horses every hour.[84] Stagecoach horses were changed every two hours.[85]

The duration of ancient battles depended largely on the stamina of the horses and the available number of horses. Confederate general Jeb Stuart is reported to have said, "He who brings the last cavalry reserve on the field during a day of battle wins."[86] During the Civil War, in the pitched battles that involved cavalry charges, fresh horses were kept in reserve at the base camp or held by a groom near the battlefield.[87] Given the need for a constant supply of fresh horses in battle, the mention of an Iron Age chariot battle that lasted continuously throughout a single day is a surprise: "The battle raged all day long, and the king remained propped up in the chariot facing Aram . . . and at dusk he died" (1 Kgs 22:35). Perhaps the "all-day" feature is merely a literary device to enhance the drama; however, if true, it is important to remember that the endurance of any active horse did not go beyond an hour or two.

### *Killing a Horse in Battle*

Prior to the advent of guns, horses were very difficult to kill. An arrow wound did not kill a horse. Rather, arrow wounds, lance wounds, and other body piercings only stimulated the horse rather than slowing it down.[88] Even when arrows hit an artery,

---

interview of Stuart Black, Olympic cross-country, event rider by K. Briggs, "Heartbreak Hill," *Hunter and Sport Horse* 10, Jan/Feb 2000, 80–90. "You can work a horse to death . . . but you can't work an ox to death. It'll just stop and say to hell with you. So which is the smarter animal?" See interview of Jochen Welsch, chief ox drover, Freeman Farm, Old Sturbridge Village, Massachusetts, by V. Klinkenborg ("Come booossss! Come booooossss!" *Smithsonian* 24, September 1993, 82–93).

84. New riders took over every 75–100 miles. The riders changed to a fresh horse every 10–15 miles. The 2,000-mile journey from St Joseph, Missouri, to Sacramento, California, could be accomplished in 10 days. See "Pony Express Information," http://www.americanwest.com/trails/pages/ponyexp1.htm (accessed 2 June 2007).

85. "The horses were changed at 'stages,' or stops, every 15–20 miles, as that was the most that could be reasonably expected from the horses"; D. A. Norris, "Stagecoaches," *History Magazine* 4, Feb/March 2003, 17–20.

86. Col. R. W. Black, *Cavalry Raids of the Civil War* (Mechanicsburg, PA: Stackpole, 2004) 16.

87. One of the longest cavalry battles of the Civil War lasted 7½ hours, with each side believing they had won (Black, *Cavalry Raids*, 29–30). In contrast, Napoleon's disastrous cavalry battle, the Battle of Waterloo, with 80 squadrons equaling 10,000 horsemen, consisted of 14 charges and lasted 2½ hours; A. Roberts, "Bravery Wasn't Enough," *MHQ* 18, Autumn 2005, 6–17.

88. An interesting comparison with killing a horse in battle is hunting a deer with a bow. A 150-lb. (68-kg) whitetail deer carries 8 pints of blood in its circulatory system. Before falling to the ground, a deer must lose 35 percent of its blood (almost three pints [1.41 L]). When frightened, deer produce high levels of B-endorphin, which supports rapid wound healing and operates as a painkiller; wounded deer routinely travel miles before falling. See http://www.myoan.net/huntingart/deer_shot_place.html.

it took hours for a horse to bleed to death.[89] In the meantime, an injured horse could actually be driven from battle and returned to camp. The warriors could then return with fresh teams of horses and replenished weapons.

Consequently, an axe between the eyes was the surest way to make a horse fall to the ground immediately. An injury of this sort was not easy to inflict on a chaotic battlefield because a horse's head was higher than the average human could reach with deadly force. From ancient chariot burials in Cyprus and Mongolia, we find that, when horses were sacrificed, presumably to accompany their humans into the netherworld, they were killed by an axe blow between the eyes. To protect this small vulnerable spot during battle, many horses wore frontlets, some of which were discovered in horse burials in Cyprus dating to the Iron Age.[90] Small cavalry horses occasionally wore bronze helmets to protect their poll, a vulnerable area located between the ears. These protective helmets, dating to the ninth century, appear designed specifically to deflect axe blows.[91]

As a practical matter, even if a warrior were close enough to a horse to deliver a fatal blow, a charioteer could run him down, or chariot fighters could spear or club him first. Even if an infantryman managed to wound one or both of a chariot's horses with a spear or knife, they possessed the power to continue long enough to knock the enemy over and allow the chariot fighters to kill him; attacking a chariot's horses was neither a simple nor a sure proposition.

If an ancient warrior really had wanted to incapacitate a horse, he would have focused on its legs, rather than shooting arrows at random. In fact, the intent to damage a horse's legs may have been the motivation behind the development of chariots with iron scythes protruding from the axles. These blades were approximately knee high on a horse, thus assuring that no horse came close to the chariot without the

---

Horses are considerably harder to kill with a bow and arrow than a deer. A 1,000-lb. (453.59 kg) horse has about 100 pints (47.31 L) of blood and can lose 30 percent of it and continue to live. Furthermore, horses have good hemostatic systems that cause blood vessels to retract back into tissue and seal themselves off. Thus, it is unlikely that a horse would bleed to death on an ancient battlefield. See D. A. Jefferson, "Blood Loss in Horses," http://www.maineequineassociates.com/ArchivedArticles/BloodLoss.htm (accessed 17 November 2007).

89. The only shot that drops a deer to the ground immediately is a spine shot. This requires shooting with a downward trajectory from a tree or deer stand, an angle not possible to duplicate on a battlefield where, at best, a mounted rider would be level with the horse. Even if a deer is hit in a vital area, such as the lungs or heart, it may still travel hundreds of yards before dropping to the ground. If hit in a non-vital area, such as the neck or gut, an injured deer could travel much farther. It is not unheard of to track wounded deer for distances of a mile or more after they are shot. Some are never found. (Personal communication with Clay O'Daniel, deer hunter, Alpharetta, GA.)

90. Crouwel, "Chariots in Iron Age Cyprus," 163, 167. Frontlets were worn by Persian cavalry horses in fifth-century battles; Xenophon, *Anabasis* 1.8 (117).

91. M. A. Littauer and J. H. Crouwel, "Ancient Iranian Horse Helmets," in *Selected Writings on Chariots, Other Early Vehicles, Riding and Harness* (ed. P. Raulwing; Culture and History of the Ancient Near East 6; Leiden: Brill, 2002) 534–44.

risk of grave danger. If the sharp blades hit at the knees or below, a horse would fall to the ground and be unable to stand again, thus completely incapacitating the battle chariot to which it was attached. There is no evidence of the use of bladed chariots in Iron Age battles in Israel, and their use in later Greek and Persian battles was generally unsuccessful.[92] The last noted use of the scythe chariot was in the Battle of Gaugamela (331) by the Persian king Darius III against Alexander the Great; it was used with questionable effect.[93]

Numerous detailed booty lists from Egypt, Assyria, and Israel reveal that a major goal of any ancient battle was actually to capture, not to kill the enemies' chariot horses. This seems sensible, given the significant investment of time necessary to train chariot horses. For the warrior, the horse's training translated unequivocally into his chances of survival and success on the battlefield. The more experienced the horse, the more reliable it was. At a minimum, readying a horse to assume even the lowliest role of "green-broke" assistant to a more seasoned chariot horse, as the Kikkuli text illustrates, required over six months of rigorous daily training.[94] However, completing the six-month training period did not mean that a horse was ready for battle. Rather, it meant that the young horse was ready to be teamed with a seasoned veteran for more advanced training. Notably, the most valued horses not only were "trained to yoke" but had seen battle and survived. The value of the seasoned war-horse may explain why the booty lists following battles almost always give horses first mention and appear to contain the exact numbers of captured horses.

Iron Age battles were won or lost depending on the best use of horses. As Xenophon succinctly states, "Success in an attempt to pursue or retreat depends on the experience of horses and their powers."[95] Traces of this fact are evident in the biblical text and can be found in advice to the Aramean king: "You must raise an army like the one you lost—*horse for horse and chariot for chariot*—so we can fight Israel on the plains" (1 Kgs 20:25; emphasis added); from the byword of the Judahite kings: "*my horses as your horses*" (1 Kgs 22:4, 2 Kgs 3:7; emphasis added); and even more emphatically in the sarcastic taunt of the Assyrian commander to the surrounding Jerusalemites (ca. 701): "I will give you two thousand horses if you can produce riders to mount them" (2 Kgs 18:23). Horses, key to the prestige of any nation, also determined the convention of warfare.

92. Crouwel, "Chariots in Iron Age Cyprus," 163, 169.

93. A. Cotterell, *Chariot: From Chariot to Tank, the Astounding Rise and Fall of the World's First War Machine* (Woodstock, NY: Overlook, 2005) 55.

94. For a detailed account of the ancient Kikkuli method of horse training used with modern-day horses, see Nyland, *The Kikkuli Method*.

95. Xenophon, *The Cavalry Commander* 5.4.

## Chapter 3

# *Horses in Iron Age Israel and Judah*

*Their land is full of horses; there is no end to their chariots.*
*(Isa 2:7)*

The availability and use of horses in Iron Age Israel and Judah are vastly underreported in modern scholarship. Some authors suggest that Israel's army was small, with chariotry and cavalry playing only minor, supporting roles.[1] Other authorities assume that horses and chariotry were inordinately expensive and question the economic ability of Israel to support a large chariotry during the Iron Age.[2] It is commonly believed that the number of horses and chariots was routinely inflated to make battles seem more important and victories more dramatic.[3] However, epigraphic, archaeological, and architectural evidence clearly supports the fact that horses were plentiful and chariots ubiquitous. Thus far, Iron Age stables have been identified by archaeologists

1. B. E. Kelle, *Ancient Israel at War: 853–586 B.C.* (Oxford: Osprey, 2007) 22. Some commentators are simply historically and archaeologically misinformed, as evidenced by M. Elat's belief that "horses were not native of Palestine and in Old Testament times they were neither bred nor kept by the inhabitants of Palestine" ("The Monarchy and the Development of Trade in Ancient Israel," in *State and Temple Economy in the Ancient Near East: Proceedings of the International Conference* [ed. E. Lipińsky; Leuven: Departement Oriëntalistiek, 1979] 527–46).

2. Israeli scholar Nadav Na'aman questions the ability of Iron Age Israel to support a chariotry requiring 4,000–6,000 horses and suggests that Ahab's 2,000 chariots (as inscribed on the Kurkh Monolith) at the Battle of Qarqar is "not historically feasible." Na'aman suggests that 200 chariots is a more realistic number, and his opinion is quoted frequently. However, his views on this topic were published before the discovery of the Tel Dan Inscription, which mentions specifically the thousands of horses in the Omride army ("Ahab's Chariot Force at the Battle of Qarqar," *Ancient Israel and Its Neighbors—Interaction and Counteraction: Collected Essays* [3 vols.; Winona Lake, IN: Eisenbrauns, 2005–6] 1:1–12; contra: L. Grabbe, "The Kingdom of Israel From Omri to the Fall of Samaria: If We Only Had the Bible . . . ," in *Ahab Agonistes: The Rise and Fall of the Omri Dynasty* (ed. L. Grabbe; Library of Hebrew Bible/ Old Testament 421; London: T. & T. Clark, 2007) 81.

3. A. F. Rainey and R. S. Notley, *The Sacred Bridge: Carta's Atlas of the Biblical World* (Jerusalem: Carta, 2006) 200.

in Israel at Megiddo, Tell Qasile, Tell Abu Hawan, and Kinrot and in Judah at Lachish, Tel Masos, Tel Malhata, Tell el-Hesi, Beth Shemesh, and Beer-sheba—all near major trade/caravan routes or militarily important roadways.[4] In fact, the numerous references in the Hebrew Bible to Israel's horses and chariotry as well as in Assyrian and Aramean texts reflect the reality of a nation in the ninth and eighth centuries committed to using the horse as its primary means of maintaining autonomy.[5]

## *Kurkh Monolith*

According to the Assyrian account on the Kurkh Monolith, Ahab, the Israelite king (869–850), fielded the largest chariotry, 2,000 strong (4,000–6,000 horses), in the Battle of Qarqar (853); Ahab's forces outnumbered even the Assyrian chariot force.[6] The Levant-centered coalition forces, with a total of 3,940 chariots (8,000–12,000 horses) and 1,900 cavalry temporarily halted the aggression of Shalmaneser III (859–824).[7] Shalmaneser returned to Assyria, although he claimed victory ("In that

4. J. S. Holladay Jr., "Stables," *OEANE* 5:69–74.

5. The prominence of the horse and chariot in warfare is emphasized in accounts of ancient Near Eastern military activity during the Bronze Age and continues in the Iron Age; Robert Drews, *The End of the Bronze Age* (Princeton: Princeton University Press, 1993) 104–34.

6. The Kurkh Monolith is a seven-foot-tall limestone, inscribed stele. The inscribed text documents the first six years of Shalmaneser's reign and primarily chronicles his military campaigns and conquests. See T. Schneider, "Did Jehu Kill His Own Family?" *BAR* 21/1 (1995) 26–33, 80–82. Online: http://www.jewishhistory.com/jh.php?id=Assyrian&content=content/did_king_jehu6 (accessed December 2007).

7. For a detailed discussion of the authenticity of these numbers or lack thereof, see M. De Odorico, *The Use of Numbers and Quantifications in the Assyrian Royal Inscriptions* (SAAS 3; Helsinki: Neo-Assyrian Text Corpus Project, 1995) 103–7. De Odorico concludes that the Kurkh Monolith numbers have been deliberately altered by a factor of 10, thus, by his estimate, making the Israelite chariotry only 200 (400 horses) and the total number of coalition horses, including cavalry and chariotry, about 1,000–1,500, depending on whether the chariots were pulled by 2 or 3 horses.

Although caution is advisable in taking ancient numbers and quantifications at face value, it is also important to acknowledge that in modern times large numbers of horses may seem especially speculative simply because they no longer play a key role in warfare and daily life. However, in 1914 C.E, the British army alone mobilized 140,000 horses in the first 10 days of World War I; E. Peplow, ed., "The Horse at War," *Encyclopedia of the Horse* (London: Chancellor, 2002) 2. In World War II, the Russian Army fielded 30 cavalry divisions, mustering a total of 1.2 million war-horses; E. H. Edwards, *The New Rider's Horse Encyclopedia* (Irvington, NY: Hydra, 2003) 29.

Furthermore, during World War I, the Desert Mounted Corps, deployed to push the Turks out of Israel in the Palestine campaign in 1917–1918, stationed 25,000 horses under the command of General Edmund Allenby at various camps in Israel; R. M. P. Preston, *The Desert Mounted Corps: An Account of the Cavalry Operations in Syria and Palestine: 1917–1918* (Boston: Houghton Mifflin, 1923) 62. At the Battle of Megiddo on September 20, 1918 C.E, often described as the last great cavalry battle, the British fielded 9,000 horses against the Turkish cavalry of about 3,000; B. Gardener, *Allenby of Arabia, Lawrence's General* (New York: Coward-McCann, 1966) 180–82. For an interesting comparison of Allenby's Battle of Megiddo with Thutmose III's (1479), see E. H. Cline, *The Battle of Armageddon: Megiddo and the Jezreel Valley from the Bronze Age to the Nuclear Age* (Ann Arbor: University of Michigan Press, 2000) 6–24.

battle, I took from them their chariots, cavalry [and] horses broken to harness"); he did not renew his invasion for four years.[8]

The fact that Shalmaneser did not establish an outpost at Qarqar or invade farther south toward Hamath and into the Levant suggests that the Assyrians perhaps *lost* the battle.[9] However, it is also possible that Shalmaneser actually seized or disabled most of the coalition's horses and chariots, and so he simply did not need to return for four years. It would take coalition forces at least that long to train and ready a sizable chariot force for battle. The immediate purpose of any battle was to immobilize the enemy by capturing their horses and destroying their chariots.

Shalmaneser is certain to have taken the horses with him after the Battle of Qarqar and not left them behind to be recaptured by his enemies. Perhaps as long as his enemies did not have enough horses to challenge him effectively in battle, Shalmaneser considered himself the victor. It is certain, however, that, even though Israel supplied the largest numbers of horses for the battle, Ahab did not send all of Israel's horses to Qarqar. Logic dictates that the broodmares, foals, and young horses in training were left safely in Israel.

The memory of the Battle of Qarqar is omitted from the biblical text.[10] Nevertheless, such a large battle could hardly have been kept secret in Israel; the deployment of troops and horses would have left a sizable vacuum. The omission of this important battle is even more puzzling when considered with the propensity of the early prophets and the later writers of Kings to extol the glory days of Israelite chariotry.[11] If Israel defeated the mighty Assyrians, one would expect the writers of Kings to refer to this very impressive victory. However, if Ahab and his regime were despicable in the view of the authors, perhaps the battle victory was intentionally ignored. Some scholars argue that, had Ahab really suffered the massive defeat recorded in the Assyrian

8. S. Yamada, *The Construction of the Assyrian Empire: A Historical Study of the Inscriptions of Shalmaneser III (859–824 B.C.) Relating to His Campaigns to the West* (Culture and History of the Ancient Near East 3; Boston: Brill, 2000) 156, 163. For a detailed discussion of the coalition's political realignments during this period, see K. L. Younger Jr., "Neo-Assyrian and Israelite History in the Ninth Century: The Role of Shalmaneser III," in *Understanding the History of Ancient Israel* (ed. H. G. M. Williamson; Proceedings of the British Academy; Oxford: Oxford University Press, 2007) 243–77.

9. M. Elat, "The Campaigns of Shalmaneser III," *IEJ* 25 (1975) 25–35. For the suggestion that the Assyrians were "stopped in their tracks," see Rainey and Notley, *Sacred Bridge*, 201.

10. Interestingly, the biblical text (Jer 46:2) does mention the Battle of Carchemish (605), a significant international battle between an Egyptian and Assyrian coalition against Babylon's Nebuchadnezzar II (605–562).

11. For example, Elisha's exclamation, "My father! My father! The chariots and horsemen of Israel!"—upon seeing the theophany of fiery chariots transporting Elijah to heaven—may have become an anthem for the Israelite military (2 Kgs 2:11–12, 6:16–17, 13:14–19). See also, M. A. Beek, "The Meaning of the Expression 'The Chariots and the Horsemen of Israel (II Kings ii 12),'" *OtSt* 17 (1972) 1–10. Concerning the Assyrian army, scholar J. N. Postgate postulates, "At any given time the size of the chariotry and the numbers of horses fluctuated, but their symbolic value never waned" ("The Assyrian Army in Zamua," *Iraq* 62 [2000] 89–108).

annals, an account of the battle would have served the anti-Ahab didactic purposes of the book of Kings.[12] Alternatively, perhaps the battle was recorded in the Hebrew text but edited out by later redactors with an anti-Ahab bias.[13] Or it may be that the accounts of Ahab as Israel's king were composed so long after the Battle of Qarqar that the memory of the encounter had faded from the historical recollection. Speculation abounds; the omission remains a mystery.

### *Tel Dan Stele*

Further evidence of the vast number of horses in Iron Age Israel is contained in the Tel Dan inscription. The Tel Dan victory stele, erected by a king of Aram (Hazael or Ben-hadad II) in the late ninth century, specifically mentions the slaughter of 70 kings, including Joram of Israel and Ahaziah of the House of David, 'who harnessed *thou*[*sands* of char]riots and *thousands* of horsemen [or: horses]" (lines 6–7; emphasis added).[14] An alternative translation by André Lemaire reads, 'And I killed two [power] ful kin[gs], who harnessed two thou[sand cha]riots and two thousand horsemen. [I killed Jo]ram of [Ahab]'.[15] Lemaire argues that the mention of 2,000 chariots confirms the size of the Israelite chariotry recorded in the Kurkh Monolith 12 years earlier and suggests that the number refers to the combined forces of Israel and Judah.[16]

The Tel Dan stele fragment was discovered in 1993 C.E.; two additional fragments were recovered in 1994 C.E. At the time of discovery, the fragments were found in secondary use incorporated into the city gate's outer courtyard. This stele provides yet another objective, "real-time" numerical account of Israel's horses and chariots.

12. W. W. Hallo and W. K. Simpson, *The Ancient Near East: A History* (ed. J. M. Blum; New York: Harcourt Brace Jovanovich, 1971) 128.

13. See also chap. 6. Na'aman attributes the omission of Qarqar from Kings to the antiquity of the events and the few sources of North Israelite origin available to the author, noting that most scholars believe the books of Kings were written 200–300 years after the recounted events of the late tenth–ninth centuries ("The Northern Kingdom in the Late Tenth–Ninth Centuries B.C.E.," in *Understanding the History of Ancient Israel* [ed. H. G. M. Williamson; Proceedings of the British Academy 143; Oxford: Oxford University Press, 2007] 400–418). For the suggestion that Ahab and his military contributions at Qarqar did not harmonize with the narrator's concept of divine will and inherent hatred of Ahab and the Omride Dynasty, see G. W. Ahlström, "The Battle at Ramoth-Gilead in 841 B.C.," in "*Wünschet Jerusalem Frieden*" (BEATAJ 13; Frankfurt am Main: Peter Lang, 1988) 157–66.

14. W. M. Schniedewind, "Tel Dan Stela: New Light on Aramaic and Jehu's Revolt," *BASOR* 302 (1996) 75-90. For the suggestion that 'horses' is a more appropriate translation for the Aramaic than 'horsemen', see I. Kottsieper, "The Tel Dan Inscription," in *Ahab Agonistes: The Rise and Fall of the Omri Dynasty* (ed. L. Grabbe; London: T. & T. Clark, 2007) 104–34. For the suggestion that the missing segment bracketed in line 6 is 'thousands' or 'myriads' of chariots, see B. Halpern, "The Stela from Dan: Epigraphic and Historical Considerations," *BASOR* 296 (1994) 63–80. Halpern notes (p. 76) that chariots usually take pride of place in victory rhetoric and that in some passages horses are treated as though they were the only military arm that mattered.

15. A. Lemaire, "The Tel Dan Stela as a Piece of Royal Historiography," *JSOT* 81 (1998) 3–14, esp. lines 6′–7′ on p. 4.

16. Ibid., 9–10.

The stele fragments are particularly interesting when considered against the context of 2 Kgs 13:1–9. There, the king of Aram-Damascus is said to have reduced Jehoahaz's army to "fifty horsemen, ten chariots, and ten thousand footmen."[17] The Tel Dan inscription, when read in accord with the biblical accounts of Israelite warfare with the Arameans, seems to confirm that Israel had a large and formidable chariotry in the ninth century that was decimated by Hazael between 842 and 810.[18] This memory of the Arameans capturing Israel's chariotry may be reflected in the 2 Kings 13 passage.

## Hebrew Bible Texts

There are numerous references in the Hebrew Bible to the horses and chariotry of Israel's enemies and to God's miraculous interventions to save the Hebrews from the certain destruction they could inflict. These references extolling enemy chariotry are particularly abundant in narratives reporting Israel's formative years. For example, the illustrious Egyptian chariotry pursued the ancient Hebrews during the exodus (Exod 14:23); Joshua captured and hamstrung the enemy horses of the Hazor coalition during the conquest (Josh 11:4); and Deborah and Barak (with divine intervention) successfully defeated the dangerous iron chariots of the Philistines during the period of the judges (Judg 4:3). By contrast, chariots and horses are not attributed textually to the Israelites during the nation's infancy.[19] In fact, the disadvantage of fighting without horses is acknowledged in Deut 20:1: "When you take the field against your enemies, and *see horses and chariots—forces larger than yours*—have no fear of them, for the Lord your God, who brought you from the land of Egypt, is with you" (emphasis added).

The scarcity of horses in narratives set before the Monarchy is also keenly apparent in the "wealth lists" of Abraham, Isaac, Jacob, and Job. Mentioned there are sheep, camels, asses, and slaves, but horses are not (Gen 12:16, 24:35, 26:14, 30:43; Job 1:3, 42:12). However, Joseph, as an Egyptian dignitary, is portrayed with a chariot and horsemen at his disposal (Gen 46:29, 47:17, 50:9).[20] Nor are caretakers for horses

17. See chap. 6 below.

18. E. Lipiński, *The Aramaeans: Their Ancient History, Culture, Religion* (OLA 100; Leuven: Peeters, 2000) 386–87.

19. Scholars debate whether Israelites or Canaanites appear in the battle scenes on the reliefs on the western face of the enclosure wall of the Cour de la Cachette at the Karnak temple dating to the time of Merneptah's rule (1212–1182). At the center of the controversy is whether the battle chariot depicted in scene 4 as identified by Yurco is Israelite or Canaanite. For an overview of these issues, see Rainey and Notley, *Sacred Bridge*, 99–100. For more in-depth analysis, see the following: A. Rainey, "Can You Name the Panel with the Israelites?" *BAR* 17/6 (1991) 56–60, 93; L. Stager, "Merenptah, Israel, and the Sea Peoples: New Light on an Old Relief," *ErIsr* 18 (1985; Avigad Volume) 56–61; F. J. Yurco, "Merneptah's Canaanite Campaign," *JARCE* 23 (1986) 189–215.

20. The reference to a viper's biting a horse's heel, causing the "rider" to tumble backward in Gen 49:17 could refer to chariotry or cavalry. Horses are instinctively afraid of snakes and often go berserk in their presence, dislodging a rider or a chariot passenger quite easily.

listed in the duty roster during the reign of David (1005–970), although the attendants for sheep, donkeys, and camels are specifically named (1 Chr 27:29–31). Interestingly, horses were not included in the memory of Israel's possession of wealth and power until the stories about Absalom and Solomon: "Absalom provided himself with a chariot, horses, and fifty outrunners" (2 Sam 15:1), and "Solomon assembled chariots and horses" (1 Kgs 10:26–29).

Furthermore, it is notable that Hebrew agricultural laws never mention horses, although stolen and wandering cattle, sheep, and donkeys are dealt with in some detail (Exod 21:32–37, 22:8–9, 23:4). The only "law" concerning horses is a prohibition against the king's acquiring many horses, especially horses from Egypt (Deut 17:16). The context of these laws suggests that horse ownership, when it was recorded, was strictly associated with the prerogative of the king, who was responsible for the state stud and chariotry.[21] However, it is important to remember that the absence of horses in the narrative does not mean they were not present. Rather, the Hebrew tradition strongly associates horse ownership with statehood and the power of the king. Notwithstanding the absence of horses in the nomadic ideals set forth in the premonarchic narratives, it is certain that, just as horses were known in Bronze Age Mari, Nuzi, Babylon, Syria, Canaan, and Egypt, they were also prevalent in the region of Israel and did not just suddenly appear in the Iron Age.

It is during the Monarchic period (1020–587) that tradition makes horses the key players in Israel's (and Judah's) fight for survival as a nation. The biblical text credits Solomon with founding Israel's chariotry, described as including 1,400 chariots and 12,000 horses (1 Kgs 10:26; 2 Chr 1:14, 9:25). Significantly, architectural accommodation for the horses, including the construction of chariot and cavalry cities throughout the country, is also mentioned (1 Kgs 9:19, 22; 2 Chr 8:6; 9:25). According to the biblical text, the infrastructure developed to support these horse compounds consisted of 4,000 stalls to contain horses, as well as a monthly rotation system directed by district officers for the feeding of the horses (1 Kgs 4:26–28). As discussed in chap. 4, below, whether the chariot and cavalry cities belonged to Solomon or to a later king continues to be debated by archaeologists and biblical scholars.

21. Interestingly, the Code of Hammurapi (ca. 1750) is also silent about horses in the Old Babylonian society, although there is extensive textual evidence of their presence in the correspondence. Apparently, the king had no more need to dictate laws about his horses than he did about his wives; they were all property of the crown. See M. T. Roth, ed., *Law Collections from Mesopotamia and Asia Minor* (2nd ed.; SBLWAW 6; Atlanta: Scholars Press, 1997) 71–142.

In Mari, King Zimri-Lim was particularly keen on acquiring fine horses and seeing to their care, as this letter to his mother (or aunt) indicates: "I keep hearing about white horses that come from Qatna [in Syria]. These are fine horses. Now then, once you listen to this tablet, in the Multicolor Yard, at the gate of [the guards?], where there is [. . . shade?], a stable should be prepared, [posts?] should be driven, so that theses horses can be quartered. Grain should be provided" (ARM 10 147 = LAPO 18 1110, pp. 290–92; translation courtesy J. M. Sasson).

Nevertheless, the numerous accounts of chariot battles between the Israelites and Arameans in the ninth century preserve the memory of an abundant supply of horses. For example, when the king of Aram loses his chariotry in battle with Ahab, his officials advise him simply to "muster for yourself an army equal to the army you lost, *horse for horse and chariot for chariot.*" Although the king returns the next spring to fight in the plains at Aphek and loses again, the implication that a chariotry could be replenished in one year suggests that the Arameans had an active breeding and training program already in place or that the king had the means to purchase fully trained replacements (1 Kgs 20:23–34). Also implied is the notion that Ahab's chariotry was quite formidable.

Likewise, from Isaiah in a late-eighth-century context, we find: "The land is full of horses; there is no limit to their chariots" (Isa 2:7). Scholars debate whether this statement references the state of affairs in Judah or Israel.[22] If addressed to Israel, it probably portrays the military buildup prior to the invasion of Judah by the Israelite-Aramean coalition during the Syro-Ephraimite war (735–732): "The presence of horses and chariots in large numbers suggest mobilization for war, and may indicate that the Aramean contingents had already taken up forward positions in Israel in preparation for the attack on Jerusalem."[23] However, if composed after this time and directed primarily at Judah, the historical context could be the impending invasion of Sennacherib in 701. If so, it is possible that the Israelite horses were evacuated to Judah during the turmoil in the North caused by the invasion of Tiglath-pileser in 732. It is equally plausible that Judah had built its own sizable defense force of chariotry in hopes of thwarting invasion by any enemy.

It is clear that there are vivid memories preserved in the Hebrew Bible of large numbers of horses in both Judah and Israel during the Iron Age. Whence came these horses? Three possibilities are logical: the horses were captured from enemies, they were bought from neighbors, or they were bred in Israel.[24]

## *Captured Horses*

One advantage of capturing enemy chariot horses was that they were already trained, experienced in battle, and ready to be redeployed. For this reason, horses were more valuable spoils of war than were men. In fact, it is quite possible that the

22. See R. Davidson, "The Interpretation of Isaiah II 6ff.," *VT* 16 (1966) 1–7; H. Wildberger, *Isaiah 1–12*: A Commentary (CC; Minneapolis: Fortress, 1990) 107–9; H. G. M. Williamson, *A Critical and Exegetical Commentary on Isaiah 1–27* (ed. G. I. Davies et al.; ICC; 3 vols.; London: T. & T. Clark, 2006) 216.

23. J. J. M. Roberts, "Isaiah 2 and the Prophet's Message to the North," *JQR* 75 (1985) 290–308.

24. A fourth possibility, that horses were acquired as tribute from subjugated enemies is more remote. However, 1 Kgs 10:25 credits Solomon with receiving annual tribute payments that included horses and mules. Additionally, after defeating the Arameans and capturing their horses, David supposedly established a garrison in Damascus and received tribute, which conceivably could have included horses (2 Sam 8:4–6).

same trained chariot horses changed owners frequently. Their useful work life, generally from ages 3 to 30, gave them longevity greater than the reigns of most rulers. However, the issue of how to handle the horses immediately after their battlefield capture posed a challenge.

According to biblical narratives, King David captured 1,000 chariots from Hadadezer on Israel's northern border. He "hamstrung" (ויעקר) all but 100 of the horses. So presumably, 1,900 horses were hamstrung (2 Sam 8:4, 1 Chr 18:4). Hamstringing horses served two distinct purposes. First, it discouraged the enemy from returning to recoup their abandoned horses. Second, it allowed the Israelites to control the captured horses (since they could not run) until corrals and accommodations were constructed for them. Indeed, in the narrative of Joshua's battle with the king of Hazor, *all* of the captured horses are hamstrung, thus indicating a memory that the ancient Israelites were at one time ill-equipped to handle and process horses (Josh 11:9). However, it appears from both accounts that another use for these captured war-horses, whether military or economic, was intended.[25]

It is important to recognize that to "hamstring" does not necessarily mean to lame permanently or to render useless.[26] Whether a horse is permanently injured by such an act depends on exactly which muscle, tendon, or ligament is cut. Generally, one thinks of the hamstring as the tendon running from the back of the knee, or "point of the hock" on a horse, to the gluteal muscle. If the surgery is performed above and behind the hock of the horse on the major tendon, the leg will collapse under the slightest weight. A cut of this sort would render any horse permanently unable to walk; destruction of the animal would be the only option. The only logical reason to do this would be to keep them alive until they could be slaughtered. Although some cultures historically had a practice of eating horses, there is no evidence that the Israelites did.[27]

25. N. K. Gottwald, *The Tribes of Yahweh: A Sociology of the Religion of Liberated Israel, 1250–1050 B.C.E.* (Maryknoll, NY: Orbis, 1979) 547.

26. The Hebrew word translated 'hamstring' is from the root עקר, which can mean 'hough' or 'cut'. It is quite possible that the horses were maimed or 'houghed' at some location other than the actual hamstring—for example, just above the hoof, where full healing would be expected.

The word עקר, meaning 'stump' in Dan 4:15, 20, and 23 is applied to the tree stump (and its roots) that were cut off but left to rejuvenate. Does this use of the word inherently suggest a principle for temporarily incapacitating something in order for it to grow back? Logically, horses were too valuable to kill because they were easily reused or sold. For limitations and prohibitions against the wanton waste of livestock in ancient Israel, see J. M. Sasson, "Ritual Wisdom? On 'Seething a Kid in Its Mother's Milk,'" in *Kein Land für sich allein: Studien zum Kulturkontakt in Kanaan, Israel/Palästina und Ebirnâri für Manfred Weippert zum 65. Geburtstag* (ed. Ulrich Hübner and Ernst Axel Knauf; OBO 186; Freiburg: Universitätsverlag / Göttingen: Vandenhoeck & Ruprecht, 2002) 294–308.

27. Xenophon mentions that horses were sacrificed and presumably eaten by the participants in Persian feasts dedicated to the Sun (*Cyropaedia* 8.3.12, 24). According to Strabo, the first-century Greek

In contrast, however, an injury to the *flexor metatarsus,* which runs down the front of the back leg from the stifle to the hock and acts as a counter to the hamstring, would allow the horse to stand normally, because the disabled tendon is non-weight bearing. A horse with an injured *flexor metatarsus* would be able to walk, albeit hesitantly. However, the horse would not be able to trot or canter until the wound healed, a process that could take weeks or months.[28] It is likely that the horses captured by David were expected to recover from this type of hamstring injury so that they might later be used—if not as chariot horses, at least as breeding stock.

As previously mentioned, hamstringing captured horses allowed their captors to keep them together and controlled. Because catching a herd of stampeding chariot horses as they sped over 40 miles per hour was impossible, hamstringing was a sensible solution to the problem. Assuming that the horses were expected to recover, we may conjecture that they were kept to sell to neighboring countries, to use as draft animals or for transportation, or to serve as the foundation of Israel's own chariotry. It is logical to suppose that the recipients of these stories may have linked the capture of horses with the subsequent development of the state stud.

References to hamstringing do not occur in other ancient texts, probably because by the Iron Age other nations had well-established chariotries and highly skilled horsemen capable of handling captured horses. Therefore, we may conclude that this tradition was specific to Israel in its developing years of statehood and was only used until an adequate horse-management program was in place.[29]

According to the biblical text, Ahab also captured the horses and chariots of the Arameans, although he did not hamstring them. Presumably, Ahab immediately integrated the horses into Israel's chariotry. The biblical text does not reveal whether Ahab captured more enemy horses the next year, when he again defeated the Aramean coalition, which returned with a reconstituted chariotry to fight in the plains (1 Kgs 20:21–34). Whether these Aramean skirmishes occurred prior to the Battle of Qarqar (853) is debated by scholars.[30] Some suggest that they reflect a later ninth-century

---

historian, the king of Armenia sent 20,000 colts to the king of Persia for sacrifice at the annual feasts of Mithra (11.14.9). Furthermore, priests guarding the tomb of Cyrus sacrificed a sheep a day and a horse every month, according to Arrian, also a Greek historian writing in the Roman period (6.29.7). For a discussion of the steppe societies and others that ate domesticated horses in the fourth and third millennium, see R. Drews, *Early Riders: The Beginnings of Mounted Warfare in Asia and Europe* (New York: Routledge, 2004) 10–30. It seems that, metaphorically, the practice of eating horses was not entirely unknown in Israel: "'At my table you will eat your fill of horses and riders, mighty men and soldiers of every kind,' declares the Sovereign Lord" (Ezek 39:20).

28. M. H. Hayes, *Veterinary Notes for Horse Owners: A Manual of Horse Medicine and Surgery, Written in Popular Language* (rev. J. F. Donald Tutt; 16th ed.; London: Paul, 1968) 393–94.

29. The only other reference to hamstringing in the biblical text is with regard to oxen in Gen 49:6.

30. K. L. Younger Jr., "Neo-Assyrian and Israelite History in the Ninth Century: The Role of Shalmaneser III," in *Understanding the History of Ancient Israel* (ed. H. G. M. Williamson; Proceedings of the British Academy; Oxford: Oxford University Press, 2007) 243–77.

political situation in the days of Jehoram, Jehoahaz, or Jehoash. If they occurred prior to the conflict at Qarqar, Ahab may have formed a substantial part of the large chariot contingent that he used against the Assyrians from the horses captured from the Arameans. In any event, it is clear that, by the mid-ninth century, Israel had a substantial chariotry and relied heavily on it in warfare.

## *Horse Trading*

During the eighth century, the Egyptians maintained extensive horse-breeding operations and engaged in horse trading at markets in the Gaza territory under the control of Tiglath-pileser III and on the eastern border of Egypt.[31] An Assyrian tablet (K. 8693) in the Kouyunjik Collection of the British Museum specifies that 50,000 horses "trained to yoke" were captured when the city (probably Memphis) was invaded by Esarhaddon ca. 670.[32] The large Nubian/Kushite horses from Egypt approached 16 hands, had broad backs, and were especially suited for chariotry.[33] Notably, Egyptian horses were a favorite of Assyrian King Sargon II (721–705), who proudly declared in an inscription that he received as tribute from Egyptian King Osorkon IV "twelve large horses of Egypt, the like of which did not exist in . . . [his] country" (716–715).[34]

The Nubian/Kushite rulers of the Twenty-Fifth Dynasty (ca. 750–650) were famous for their love of horses. For example, the Victory Stele of King Piye (= Piankhy 747–716) prominently depicts a foal being presented as tribute and records the king's anger at the enemy's treatment of horses during the siege:

> His majesty proceeded to the stable of the horses and the quarters of the foals. When he saw they had been [left] to hunger he said: "I swear, as Re loves me, as my nose is refreshed by life: that my horses were made to hunger pains me more than any other crime you committed in your recklessness![35]

King Piye proceeded to subdue other local rulers, inspect their stables, and select their finest horses as booty.

At El Kurru, the royal burial grounds for the Kushite kings, excavators uncovered a horse cemetery with 24 graves. The horses were buried standing up and decorated in fine cloth, silver beads, and elaborately carved trappings, as well as shell necklaces.

31. S. Dalley, "Ancient Mesopotamian Military Organization," in *CANE*, 1:413–22; L. A. Heidorn, "The Horses of Kush," *JNES* 56 (1997) 105–14.

32. W. G. Lambert, "Booty from Egypt," *JJS* 33 (1982) 61–70.

33. The large chariot horses seen on Assyrian reliefs dating to the eighth century and later are almost certainly Egyptian. The bulky body type of Sennacherib's (704–681) horses differs significantly from the smaller, lighter horses used by Shalmaneser III (858–824), as depicted on the ninth-century Balawat Gate; Y. Yadin, *The Art of Warfare in Biblical Lands in Light of Archaeological Study* (trans. M. Pearlman; 2 vols.; New York: McGraw-Hill, 1963) 2:403, 432.

34. A. Fuchs, *Die Annalen des Jahres 711 v. Chr.: Nach Prismenfragementen aus Ninive und Assur* (ed. R. M. Whiting; SAAS 8; Helsinki: Neo-Assyrian Text Corpus Project, 1998) lines 8–11.

35. "Victory Stela of King Piye" (Lichtheim, *AEL*, 3:72).

The inscribed objects buried with the horses identify them as the chariot horses of Kings Piye, Shabako, Shebitku, and Tanwetamain, the four principal kings buried there.[36] The abundance of intricately constructed trappings, carved beads, and other jewelry connects the horses with royalty and indicates that they were especially prized by their owners.[37]

Given the widespread breeding operations under the Nubian/Kushite kings, it is possible that Egypt had a surplus of horses available for trade during the ninth and eighth centuries.[38] The biblical text preserves a strong memory of horse-trading with Egypt and suggests that Egyptian horses were readily available and easy to acquire: "The king, moreover, must not acquire great numbers of horses for himself or make the people return to Egypt to get more of them, for the Lord has told you, 'You are not to go back that way again'" (Deut 17:16). The necessity of voicing the prohibition against trading horses with Egypt presumes that activity of this sort was commonplace at some point in Israelite history.

Similarly, when faced with the certainty of an imminent Assyrian invasion of Judah, the eighth-century prophet Isaiah issued a particularly strong invective against trusting in Egypt for horses:

> Ha! Those who go down to Egypt for help and rely upon horses! They have put their trust in abundance of chariots, in vast numbers of riders, and they have not turned to the Holy One of Israel. . . . For the Egyptians are men, not God, and their horses are flesh, not spirit. (Isa 31:1, 3)

It is unclear whether Israel hired the Egyptian chariotry and charioteers as mercenary units or merely leased the horses and chariots for use by its own officers.[39] Regardless, it is apparent that King Hezekiah (715–686) or at least some of his influential advisers considered the acquisition of Egyptian horses and chariots as a viable political/military option, thus provoking Isaiah's stern warning.

According to 2 Kgs 18:23–24, by the late eighth century, Judah's reliance on Egypt for horses was a joking matter to the Assyrian army officer dictating terms of surrender to Jerusalem: "I'll give you two thousand horses if you can produce riders to

36. See Dows Dunham, *The Royal Cemeteries of Kush*, vol. 1: *El Kurru* (5 vols.; Cambridge: Harvard University Press, 1950) 110, pls. 28–29.

37. Ibid., pls. 42–43 and 49.

38. "Zerah the Cushite marched out against them with an army of a thousand thousand [*sic*] and 300 chariots. When he reached Mareshah, Asa [king of Judah, 913–873] confronted him, and the battle lines were drawn in the valley of Zephat by Mareshah" (2 Chr 14:9). Additionally, "The Cushites and Lybians were a mighty army with chariots and horsemen in very great numbers, yet because you relied on the Lord He delivered them into your hands" (2 Chr 16:8). These accounts, although written much later than the events they record, may suggest a memory that horses were trafficked into Judah from Egypt in the early ninth century or later.

39. See also Ezek 17:15: "But [that prince] rebelled against him and sent his envoys to Egypt to get horses and a large army." The hiring of mercenary chariotry from the Arameans is also recalled as a practice during the United Monarchy (1 Chr 19:7).

mount them. So how could you refuse anything even to the deputy of one of my master's lesser servants, *relying on Egypt for chariots and horsemen*?" (emphasis added). The memory of buying horses from Egypt is also preserved in the Solomon narratives, which explicitly state that Solomon purchased chariot horses from Egypt (1 Kgs 10:28, 2 Chr 9:28).

In addition to the acquisition of horses from Egypt, the biblical text mentions the purchase of horses from Kue (that is, Cilicia in the southeast of Anatolia) by Solomon's royal merchants at a fixed price (1 Kgs 10:28, 2 Chr 1:16). Perhaps this reflects either a volume discount or the fact that these horses were not yet trained to the yoke. A reference to the acquisition of horses for military use in Kue dating to the beginning of the seventh century was discovered on the portal orthostat at the Iron Age fortification of Karatepe (ancient Cilicia, modern Turkey).[40] There is also a reference in Ezek 27:14 to an active horse trade between Tyre and Beth Togarmah (usually associated with modern Armenia). In addition, as discussed in chap. 5 below, in the first half of the eighth century, Megiddo functioned as a major chariot training center and also may have been a marketplace for buying and selling trained horses.[41]

### *Horse Prices*

For over 1,000 years, trained chariot horses were the single most expensive commodity in the ancient Near East. One of the earliest horse-trading references is from a third-millennium text from Ebla (Tell Mardikh, Syria). This reference concerns the exchange between Yirkab-Damu of Ebla and Zizi of Hamazi of 2 boxwood wagons and 10 ropes (that is, chariots and harnesses) for "fine quality" equids.[42] From Mari during the reign of Yasmah-Addu (ca. 1795), an especially interesting letter sent from Iši-Addu, the king of Qatna (city in Syria), to Išme-Dagan of Ekallatum discusses the high value of horses:

> This matter is not for discussion; yet I must say it now and vent my feelings. You are the great king. When you placed a request with me for 2 horses, I indeed had them conveyed to you. But you, you sent me [just] 20 pounds of tin. Without doubt, when you sent this paltry amount of tin, you were not seeking to be honorable with me. Had you planned sending nothing at all—by the god of my father!—would I be angry! Among us in Qatna, the value of such horses is 600 shekels [= 10 pounds] of silver. But you sent me just 20 pounds of tin! What would anyone hearing this say?[43]

40. For translation and commentary, see K. L. Younger Jr., "The Azatiwada Inscription," *COS*, 2:148–50.

41. D. O. Cantrell and I. Finkelstein, "A Kingdom for a Horse: The Megiddo Stables and Eighth Century Israel," in Megiddo *IV: The 1998–2000 Seasons* (ed. I. Finkelstein, D. Ussishkin, and B. Halpern; 2 vols.; Monograph Series 24; Tel Aviv: Tel Aviv Institute of Archaeology, 2006) 2:643–65.

42. P. Michalowski, *Letters from Early Mesopotamia* (ed. Erica Reiner; Atlanta: Scholars Press, 1993) 13–14.

43. ARM 5 20 = LAPO 16 256, pp. 403–5, translation courtesy of J. M. Sasson.

The request for two horses probably indicates a trained chariot team. Even if the 300 shekels per horse is inflated tenfold, it is an impressive number, making horses the most expensive commodity in Old Babylonian society. There is evidence from Mari that a horse was worth 5 mina of silver, the equivalent of 30 slaves or 500 sheep.[44]

Mules were also highly valued and in some cases were preferred over horses, as is evident in a letter from one of Zimri-Lim's (1779–1757) advisers concerning plans for his ceremonial entrance into a city:

> "Since you are the king of the nomads and you are, secondly, king of Akkad [land], my lord ought not ride horses; rather, it is upon a palanquin or mules that my lord ought to ride, and in this way he can give honor to his majesty." This is what I told my lord.[45]

Whether this advice was given to aid the king in impressing the local mule-riding nomads or because mules are typically calm natured and, therefore, inherently safer and more dignified to ride (or drive) in parade settings is not certain.[46] The high value of mules is also attested in Old Hittite legal texts (1650–1500), where the price of a mule is listed as 40 shekels, compared with a draft horse at 20 shekels, a plow ox for 12 shekels, or a sheep for 1 shekel.[47] Mules are hybrids produced by breeding female horses to male donkeys and, therefore, infertile. This means that mules were valued solely for riding or pulling chariots and not for their reproductive potential. In general, mules are somewhat more surefooted than horses and may be stronger and have more endurance.[48] However, they are decidedly slower than horses and would not be a good choice for chariotry if the enemy possessed horses.

Records from the Ugarit Palace archives (dated from the fourteenth to thirteenth century) show a range of horse prices from 30 shekels for a mare to 200 shekels for a horse sold to the king of Ugarit by a man from Carchemish.[49] By comparison, the

44. J. M. Sasson, *The Military Establishments at Mari* (Studia Pohl 3; Rome: Pontifical Biblical Institute, 1969) 13.

45. ARM 6 76 = LAPO 17 732, pp. 484–88, translation courtesy of J. M. Sasson.

46. For an interesting article on the value of mules as mounts for rulers and dignitaries, see C. Michel, "The *perdum*-Mule, a Mount for Distinguished Persons in Mesopotamia during the First Half of the Second Millennium B.C.," in *PECUS: Man and Animal in Antiquity—Proceedings of the Conference at the Swedish Institute in Rome, September 9–12, 2002* (ed. B. S. Frizell; Swedish Institute in Rome Projects and Seminars 1; Rome: Swedish Institute, 2004) 190–200. For a discussion about the role of mules in ancient Egypt, see K. M. Hansen, "'Mules' of the 18th Dynasty," in *Ancient Egypt, the Aegean, and the Near East: Studies in Honour of Martha Rhoades Bell* (ed. J. Phillips et al.; 2 vols.; San Antonio, TX: Van Sicklen, 1997) 1:219–26.

47. Roth, *Law Collections,* 235.

48. Mules could cover up to 50 mi. (80 km) per day which was on average considerably farther than oxen traveling at 2 mph (3.21 km), hampering the speed of an army; see John Shean, "Hannibal's Mules: The Logistical Limitations of Hannibal's Army and the Battle of Cannae, 216 B.C.," *Historia* 45 (1996) 159–87.

49. C. Virolleaud, *Textes accadiens et hourrites des archives est, ouest et centrales* (Le Palais royal d'Ugarit 3; Mission de Ras Shamra 6; Paris: Imprimerie nationale, 1959) 180.

price of one ox was 10 to 17 shekels, one donkey cost 10 shekels, and the price of one sheep was about 1 shekel in Ugarit.[50] It seems that horses were more expensive in Ugarit than in Nuzi and Hattai.

During the Bronze Age, horses trained to yoke were among royal dowry items and gifts. For example, Tušratta, king of Mitanni (1380), sent to Amenhotep III (1390–1353) as dowry for his sister: four beautiful horses "that run (swiftly)," a chariot overlaid with 320 shekels of gold, and exquisitely decorated horse trappings (bridles, harnesses, and necklaces) overlaid in over 100 shekels of gold.[51] When Ashur-uballit, king of Assyria (1365–1330), initiated diplomatic relations with Egypt, he sent a gift to Akhenaten (1358–1340) of a "fine chariot and two horses."[52] Soon after he received Egyptian emissaries in his court, the Assyrian king sent more special horses to the pharaoh:

> I have dispatched to you as a peace offering a beautiful royal chariot (from among those) that I myself drive and two white horses that I likewise drive (myself); one chariot without a team of horses; and one seal of beautiful lapis lazuli.[53]

Chariot horses so were highly esteemed in Egypt during the Amarna period that the salutations in correspondence from other kings typically included best wishes for the pharaoh's horses and chariots.[54]

Throughout the Iron Age, horses remained very expensive. The biblical text, which records Solomon's purchase of horses from Egypt and Kue, sets the price at 150 shekels apiece (1 Kgs 10:28–29, 2 Chr 1:16–17).[55] With the exception of a solitary reference to David's purchase of a threshing floor and oxen for 50 shekels in 2 Sam 24:24, no prices for the purchase of animals are found in the biblical text. Some scholars believe that 150 shekels for one horse indicates a horse of superior quality and advanced training. However, this price is not extreme when compared with other documented prices for horses between 1780 and 540. In fact, a document from the palace archives of Nabonidus (555–539), king of Babylon, notes that Nabonidus paid 230 shekels for a horse of the "best quality."[56] Given this historical framework, the 150-shekel price

50. M. Heltzer, *Goods, Prices and the Organization of Trade in Ugarit: Marketing and Transportation in the Eastern Mediterranean in the Second Half of the II Millennium* B.C.E. (Wiesbaden: Reichert, 1978) 74–75, 86.

51. EA 22:1–4, found in W. L. Moran, ed., *The Amarna Letters* (translation of Tell el-Amarna Tablets; Baltimore, MD: Johns Hopkins University Press, 1992) 51.

52. A. K. Grayson, *Assyrian Royal Inscriptions* (2 vols.; Wiesbaden: Harrassowitz, 1972) 1:48, no. 310.

53. Ibid., supra p. 48, no. 11, 315.

54. See EA 1–3, 5–12, 17, 19–21, 23–24, 27–29, 33–35, 37–39, 41–42 in Moran, *The Amarna Letters*.

55. The price of Solomon's horses imported from Egypt is given as 50 shekels in the LXX translation.

56. Y. Ikeda, "Solomon's Trade in Horses and Chariots in Its International Setting," in *Studies in the Period of David and Solomon and Other Essays: Papers Read at the International Symposium for Biblical Studies, Tokyo, 5-7 December, 1979* (ed. T. Ishida; Winona Lake, IN: Eisenbrauns, 1982) 215–38.

mentioned in the biblical text, presumably for a high-quality horse with reproductive potential, seems in keeping with the typical prices paid by royalty for horses.[57]

Clearly, the high price of horses provided an economic incentive to breed rather than buy horses. The effect of such an enterprise is evident from the low price of mares (30–50 shekels) recorded in Nuzi and Ugaritic texts dating to the Late Bronze Age.[58] Of course, then as now, quality breeding stock and highly trained horses could always command premium prices.[59] It is likely that the horse prices in historical records are the record of exceptional sales for superior stallions and mares or especially well-trained teams; ordinary horses were probably traded for less.

## *Horse Breeding*

It is highly probable that Israel began its own breeding program for chariot horses; prudence demanded it.[60] As necessary strategic weapons, chariot horses were too important merely to depend upon capturing them from enemies. From an economic standpoint, raising horses for the military was far cheaper than buying them. A horse-breeding program, such as the example provided in table 3.1, would also have been fairly simple to accomplish, for in less than 12 years only 100 mares and 10 stallions could yield over 1,000 trained chariot horses.[61]

57. Even as late as the mid-fourth century, the average price of a cavalry horse in Athens was around 500 drachmas, more than the average price of a house at the time. I. G. Spence, *The Cavalry of Classical Greece* (Oxford: Clarendon, 1993) 272–79.

58. Ikeda, "Solomon's Trade in Horses and Chariots," 229–30.

59. As an indication of the value of premier horses today, at the 15-day annual Keeneland Breeding Stock Sale for racehorses in Lexington, KY, ending November 19, 2007, the average price for the 3,381 horses sold was $100,821, the second-highest average in its history. Thirty-nine horses sold for over 1 million dollars. Also setting a new record was the all-time highest price paid for a broodmare, Playful Act: $10,500,000 paid by Sheikh Mohammed bin Rashid al Maktoum; G. A. Hall, "Playful Act's Record Sale Serious Business," *The Courier-Journal*, Nov. 6, 2007.

60. Some familiarity with horse breeding in Judah is obvious from Jer 5:8: "They were well-fed, lusty stallions, each neighing at another's wife." Actually, translating מיזנים as 'well-fed' makes no sense from a horse-breeding perspective, because "well-fed" horses neigh less, not more. However, it is typical for stallions to neigh loudly when in the vicinity of mares in season. The better translation is 'they were "well-endowed," lusty stallions . . .'. The LXX reads, 'They became as wanton horses: they neighed each one after his neighbor's wife'.

61. Not all stallions are used for breeding. Only those with the best physical and mental characteristics are selected as studs. In their natural environment, horses organize into "harems" or breeding bands (usually 10 or more mares and 1 stud) and "bachelor bands" composed of all male nonbreeding horses. Although today some stallions are used for both showing and breeding (good show records equate to higher stud fees), it is more customary to use a stallion for a single purpose during a single time frame; E. Squires, *Understanding the Stallion* (Lexington: Eclipse, 1999) 86.

When a horse is "standing at stud" it usually has retired from its show/work career, at least temporarily, and is devoting its time to breeding mares. Although studs still need to be exercised, the stresses of a training regimen necessary for competition combined with breeding two or three mares a day are

*Table 3.1. Projected Horse-Breeding Program*

| *Year* | *Mares* | *Foals* | *Fillies* | *Colts* | *Chariot Horses Available* |
|---|---|---|---|---|---|
| 1 | 100 | 100 | 50 | 50 | X |
| 2 | 100 | 100 | 50 | 50 | X |
| 3 | 100 | 100 | 50 | 50 | 50 |
| 4 | 150 | 150 | 75 | 75 | 100 |
| 5 | 200 | 200 | 100 | 100 | 150 |
| 6 | 250 | 250 | 125 | 125 | 225 |
| 7 | 325 | 325 | 162 | 162 | 325 |
| 8 | 425 | 425 | 212 | 212 | 450 |
| 9 | 550 | 550 | 275 | 275 | 612 |
| 10 | 712 | 712 | 356 | 356 | 824 |
| 11 | 924 | 924 | 462 | 462 | 1,099 |
| 12 | 1,199 | 1,199 | 599 | 599 | 1,455 |

These calculations, which are only for illustrative purposes, assume that, during the first 3 years of the breeding program, 100 mares have 100 foals, of which half are fillies (female) and half are colts (male).[62] The average gestation period for a mare is 336 days.[63] Mares typically go into heat when the daylight begins to lengthen in the spring.[64] The average period of heat is 4–6 days and occurs at intervals of 21 days, unless the mare is bred and pregnant. Mature stallions of draft-type breeds (similar to the Kushite horses) may breed up to 100 mares per season, although 50–70 is the

considered too taxing; overworking a breeding stallion may lower his sperm count, resulting in failed conception (ibid.).

62. "On the average, and when considering a large population, approximately equal numbers of males and females are born in all common species of animals"; M. E. Ensminger, *Stockman's Handbook Digest* (Danville, IL: Interstate, 1992) 73. Colts must be separated from fillies before they reach puberty at 15–24 months to ensure that they do not impregnate the young fillies; Squires, *Understanding the Stallion*, 19.

63. The 11-month gestation period is followed by a "foal heat" that occurs 7–10 days after the birth of the foal, during which time the mare is typically "bred-back" to ensure that she will bear a foal at the appropriate time the following year; C. M. Schweizer, *Understanding the Broodmare* (Lexington, KY: Eclipse, 1998) 120.

64. Mares, affected by length of daylight, temperature, and other factors, are seasonal breeders. In the northern hemisphere, their natural mating season begins in April and continues through September. Likewise, stallions are "long-day breeders," with peak sperm output occurring from April to June (Squires, *Understanding the Stallion*, 6; J. M. Giffin, and T. Gore, *Horse Owner's Veterinary Handbook* (2nd ed.; New York: Howell, 1998) 341–42.

norm for other breeds.[65] Although table 3.1 only depicts a 12-year period, a stallion may remain a reliable and vigorous breeder for 20–25 years, which is slightly longer than a mare's reproductive life.[66] For simplicity's sake, these calculations assume that all horses survive, but even with a 20 percent fatality rate overall, 1,000 trained chariot horses can be ready for battle use within a decade.[67]

During Ahab's 22-year reign prior to the Battle of Qarqar (853), he could easily have raised and trained 2,000 chariot horses. Even if Ahab lost 4,000 trained chariot horses to Shalmaneser III at the Battle of Qarqar, if as few as 400 broodmares were left in Israel, the entire force could have been replenished in 10 years. This assumes adequate pasturage and grain supplies, of course, which may not have always been available. According to the biblical text, during a severe drought in Samaria lasting three years, Ahab summoned his palace administrator, Obadiah, and gave orders to save the horses. The task was so important that Ahab personally scouted for water and grass.

> Go through the land, to all the springs and valleys. Maybe we can find some grass to keep horses and mules alive so we will not have to kill any of our animals. So they divided the land they were to cover, Ahab going in one direction and Obadiah in another. (1 Kgs 18:5–6)

This passage may suggest that during droughts surplus horses were exterminated; if so, saving the broodmares and best stallions to replenish the herd when weather conditions were more favorable would be a priority.

## *Size and Breed of Horses*

Assessing the size and breed of horses available in Iron Age Israel is somewhat problematic when attempted in comparison with modern breeds, which range in size from large draft breeds such as Clydesdales and Shires (16.2–18 hands) to ponies (under 14.2 hands) to miniature horses used as pets (34 inches).[68] However, to understand

65. If stallions are bred too frequently, their semen count becomes too low to produce offspring. Furthermore, breeding season can be extremely tiring for stallions; if too many mares are presented, they may lose interest in the process and refuse to participate. This is why, when stallions begin breeding at age 2, they are typically limited to 10–15 mares during their first season.

66. Ensminger, *Stockman's Handbook*, 68–69.

67. "The annual mortality rates for feral horses in the United States under one year of age ranges from about 8–14%" (S. L. Olsen, "Early Horse Domestications: Weighing the Evidence," in *Horses and Humans: The Evolution of Human Equine Relationships* [ed. S. L. Olsen et al.; BAR International Series 1560; Oxford: Archaeopress, 2006] 81–113).

68. A "hand" is the approximate width of a human hand, 4 in. (10 cm). Horses are measured from the highest point of the withers (shoulder) to the ground in a perpendicular line. The measurement is tabulated in hands and inches; for example, 14.2hh (hands high) means 58 in. from the withers to the ground (18hh = 72 in.); S. McBane, *The Illustrated Encyclopedia of Horse Breeds* (ed. C. Marriot; Edison, NJ: Wellfleet, 1997) 13.

better the differences in types and uses of horses in the ancient world, we will find it helpful to consider the modern context. The common designation of horses as "cold-bloods" (heavy, draft-type horses from cold European climates—for example, Shires) or "hot-bloods" (light, slender-legged, sleek horses from southern warm climates—for example, Arabians) or "warm-bloods" (a mixture with the best characteristics of both—for example, Hanoverians) is a reflection of the species' adaptation to different climatic regions and has no particular scientific significance.[69] However, the classifications are useful because they are generally indicative of size and demeanor. For example, draft horses (cold-blooded) generally have deep, muscular bodies set on short, strong legs and are slow, methodical movers with calm temperaments. They can pull or carry enormous loads and may weigh up to a ton. Hot-blooded horses, such as Arabians, generally have alert, more excitable temperaments and are much leaner in appearance. They are faster and more nimble movers (typically weighing less than 1,000 lb. [453 kg]) and tolerate heat well. Although during the Medieval period the stocky, cold-blooded type horses were used by heavily armored knights as "chargers," they typically are not considered good riding horses for activities that require great speed and agility. Today's riders prefer European sport horses, that is, "warm-bloods" (for example: Hanoverians, Oldenburgs, and Dutch Warmbloods), for Olympic competitions in show-jumping and dressage because they are medium-build horses possessing great athletic ability, speed, stamina, and relatively calm temperaments.[70]

Of course, these actual classifications did not exist as such during the Iron Age, but interestingly, the idea behind them did.[71] We know from their meticulous record keeping and the depictions on palace walls and reliefs that the Assyrians designated horses by breeds, which were based on their size and utility, and often by color, which may or may not have been breed specific.[72] In Assyria, incoming horses

69. E. Peplow, ed., "The Pre-Domestic Horse," *Encyclopedia of the Horse* (London: Chancellor, 2002) 10.

70. Judy Goldman, "Warmbloods: A Hot Topic," *Showlife* 3/2 (2008) 10–11, 18–19.

71. A study of Eighteenth-Dynasty Egyptian horses reveals two distinct types reflected in the art of the second half of the fifteenth century. Catherine Rommelaere identifies two "breeds" of Egyptian horse and suggests that the slender, elongated type of horse is similar to the Akhal Temple horse of Central Asia, and the more stocky, short horse is comparable to the Arabian-Parsang. However, it is confusing to use such particular breed terminology to describe ancient horses, because the breed nomenclature is only recently developed. It is more productive to note the general size and conformation of ancient horses to determine their use as being appropriate for riding or chariot driving than to speculate which modern breeds they appear to resemble. C. Rommelaere, *Les chevaux du Nouvel Empire égyptien: Origines, races, harnachement* (Brussels: Connaissance de l'Égypte ancienne, 1991) 53–61.

72. For an insightful work on the depictions of different breeds of horses in Assyrian art, see P. Albenda, "Horses of Different Breeds: Observations in Assyrian Art," in *Nomades et sédentaires dans le Proche-Orient ancien: Compte rendu de la XLVI[e] Rencontre assyriologique internationale, Paris, 10–13 juillet 2000* (ed. C. Nicolle; Amurru 3; Paris: Éditions Recherche sur les civilisations, 2004) 321–34. See also F. M. Fales and J. N. Postgate, *Imperial Administrative Records, Part II: Provincial and Military Administration* (ed. Simo Parpola et al.; SAA 11; Helsinki: Helsinki University Press, 1995) 69, no. 113.

were inspected by the "mayor" (better: "inspector") of the Temple of Nabû at Calah, whose job was to report to the king on their origin, breed, and eventual disposition as riding horses (*pēthallu*) or yoke horses (*ša nīri*).[73] During the reign of Esarhaddon (681–669), at least three breeds were identified: Mesean, Egyptian, and Kushite. All were apparently used as chariot horses, although the Mesean horses from cities or provinces to the east of Assyria (Iran) were rarer and often dedicated to the service of the gods. [74] The Mesean horse may be the original "Caspian" horse from the area near the Caspian Sea.[75]

The most important distinction is between the riding horses and the yoke horses. Although is not impossible to train any size or breed of horse to pull a chariot, the larger, more substantial horses are typically better suited for pulling loads. Riding horses typically are smaller with a more narrow back that enables the rider to "ride with a longer leg" and grip more securely. Given Israel's strategic location between Assyria and Egypt, it is logical to assume that all of these horse breeds, especially the Egyptian and Kushite chariot horses, were present there as well. However, an intriguing passage from Habakkuk (attributed to the late seventh century) indicates that the Babylonian cavalry was mounted on horses renowned for their speed and perhaps uncommon in Israel: "Their horses are swifter than leopards, fiercer than wolves at dusk. Their cavalry gallops headlong; their horsemen come from afar" (Hab 1:8, NIV). This passage may also be an indication of the cataclysmic change from chariotry to mounted warfare.[76]

## *Economic and Practical Considerations*

The expense of rearing horses depends on the availability of rich pasture land, grain, and water, of which Iron Age Israel had plenty.[77] Although horses can consume 10 times more grain than humans on a daily basis, it is only when horses are confined for training or in preparation for war that they need to be "grained up"; at other times,

73. S. W. Cole and P. Machinist, *Letters from Priests to the Kings Esarhaddon and Assurbanipal* (ed. S. Parpola et al.; SAA 13; Helsinki: Helsinki University Press, 1998) 27–28.

74. J. N. Postgate, *Taxation and Conscription in the Assyrian Empire* (Studia Pohl: Series Maior 3; Rome: Pontifical Biblical Institute, 1974) 7–13.

75. The Caspian horse, only 10 to 12 hands, is the second-oldest breed of horse after the Przewalski horse and is thought to be an ancestor of the modern Arabian breed; Peplow, "The Pre-Domestic Horse." See also J. Clutton-Brock, *Horse Power: A History of the Horse and the Donkey in Human Societies* (Cambridge: Harvard University Press, 1992) 30–33.

76. See chap. 7.

77. For the statistical capacity of the huge grain silos in Iron Age Israel and the extent to which they indicate a pattern of accumulation of agricultural surplus, see J. S. Holladay Jr., "The Kingdoms of Israel and Judah: Political and Economic Centralization in the Iron IIA–B (ca. 1000–750 B.C.E.)," in *The Archaeology of Society in the Holy Land* (ed. T. E. Levy; London: Leicester University Press, 1998) 368–98. For specific water sources in the Lower Galilee, see Zvi Gal, *Lower Galilee during the Iron Age* (trans. Marcia Reines Josephy; American Schools of Oriental Research Dissertation Series 8; Winona Lake, IN: Eisenbrauns, 1992) 4–7.

pasturage is adequate. Horses in training should receive grain two or three times per day. The modern formula is 1.5 lb. (.68 kg) of grain and 1 lb. (.45 kg) of hay per 100 lb. (45 kg) live weight of horse. Ideally, a 1,000-lb. (454-kg) horse eats 12–15 lbs. (5.4–6.8 kg) of grain per day and 10 lb. (4.5 kg) of hay.[78] While broodmares benefit from grain during late pregnancy and while nursing, most will produce normal, healthy offspring without it.

As mentioned previously, some scholars argue that the prohibitive expense of a chariotry would have limited its use in Egyptian and other Bronze Age armies to a supportive role.[79] Similar arguments are made regarding Israel.[80] However, the ongoing cost of maintaining a chariotry was directly related to the grain harvest.[81] In Israel, where the horses were kept at Jezreel and Megiddo, literally adjacent to the abundant grain sources from the Megiddo plain and the Jezreel Valley, the cost of feed and its distribution was minimized.[82] In years of surplus grain supplies, the cost of feeding the horses was negligible for a state-run business with an effective distribution system. True, horses eat more than people do; but when there is a grain surplus, there is enough for both. The cost is merely that of labor, and at a state-run military outpost this was nil. The work was performed by soldiers and grooms, often children, as part of their routinely assigned duties.[83]

Neither stables nor training areas are necessary to breed and raise horses during their early years.[84] It would be incorrect to assume that all chariot horses required stables with individual feeding troughs such as those found at Megiddo, made of carved ashlar and dating to the first half of the eighth century.[85] Horses can be fed

78. Ensminger, *Stockman's Handbook,* 217. It is also important to consider that ancient grains, undiluted by hybridization, may have been inherently more nutritious than modern varieties, and, therefore, smaller amounts were needed to satisfy the horse's caloric intake requirements.

79. A. R. Schulman, "Military Organization in Pharaonic Egypt," in *CANE,* 1:301. See also, Drews, *The End of the Bronze Age,* 106–13.

80. Naʾaman, "Ahab's Chariot Force," 6–10.

81. Even if the generous estimate by Stuart Piggott that 8–10 acres of good grain-land would have been required to feed one team of chariot horses for a year is applied to only the 25-square-mile area in and around Megiddo, there is ample acreage to support over 3,000 horses ("Horse and Chariot: The Price of Prestige" [paper presented at the Seventh International Congress of Celtic Studies, Oxford, 10 July 1983] 25–30).

82. Rainey and Notley refer to this area as the "great breadbasket of the kingdom" (*Sacred Bridge,* 176).

83. For artistic depictions of horses being attended by young grooms from Egypt, see, Yadin, *Art of Warfare,* 1:236–37. The Assyrian annual muster included recruiting young boys for grooms; F. M. Fales, "Preparing for War in Assyria," in *Économie antique: La guerre dans les économies antiques* (ed. A. Andreau, P. Briant, and R. Descat; Entretiens d'archéologie et d'histoire 5; Saint-Bertrand-de-Comminges: Musée archéologique departemental, 2000) 35–62.

84. For a discussion of the extensive equine-management programs in the third millennium at Mari and Chagar Bazar, see F. van Koppen, "Equids in Mari and Chagar Bazar," *AoF* 29 (2002) 19–30.

85. As discussed in chap. 5 below, the Megiddo stables and training complex evidences a highly sophisticated, elaborately planned horse-management program.

their grain atop their hay (as is done at horse shows today) or in shallow bowls or feed bags, rather than in fixed troughs. Mares and foals need only pasture, water, ample space to run, and protection from predators. All of these requirements were readily available in Iron Age Israel. Highly suitable breeding grounds existed in the Gilboa Mountains near the horse compounds at Megiddo and Jezreel, and in the Hula Valley between Hazor and Dan.[86] Both locales were rich in water and grass and lay in proximity to the large Iron Age granaries.

Water availability, of course, is essential to raising horses. Horses drink 8–10 gallons of water per day under normal conditions, although this amount may double in hot climates.[87] Most ancient cities and villages were located near streams to provide adequate water for humans and livestock.[88] The watering of horses may also be accomplished from water caught in cisterns or drawn from wells. Horses are easily taught to drink from bottles in modern times and surely drank from goatskins when necessary in ancient times. Although the lack of water may have become an issue on long campaigns, as happened in the account of the invasion into Moab led by King Joram of Israel and King Jehoshaphat of Judah (2 Kgs 3:9), typically the armies could take advantage of the abundant springs and streams and plan accordingly.[89] Sites identified with stables in Judah, such as Malhata, Masos, Beer-sheba, and Tell el-Hesi were located near Wadi es-Saba and Wadi el-Hesi, which had adequate fresh water. Water supplies were plenteous in the North near the sites of Megiddo, Jezreel, Hazor, and Dan.[90] Even during the severe drought mentioned in 1 Kings 18 previously referenced, the implication is that the springs remained viable and were sufficient to produce grass for the horses.

86. Today, only 20 miles from Megiddo in Sarona, Israel, the Double K Ranch is a thriving breeding and training center for quarter horses.

87. See P. MacGregor-Morris, ed., *The Book of the Horse* (New York: Exeter, 1987) 132. A horse can sweat away 50 lbs. of fluid on a 37-mi. (60-km) ride; H. Schott, "Do Electrolytes Really Help during Endurance Exercises?" *USA Equestrian* 67 (2002) 64–65. However, the daily-minimum water ration for the horses on the British campaign in Palestine in World War I was only 5 gallons, due to the limited haulage capacity of the supply trains, composed of over 20,000 camels and 12,000 donkeys. By comparison, the soldier's daily water ration was 1 gallon, of which 2 pints were intended for drinking; Marquess of Anglesey, *A History of the British Cavalry, 1816–1919*, vol. 6: *1914–1918: Mesopotamia* (London: Pen and Sword, 1995) 35, 207.

88. J. P. Oleson, "Water Works," *ABD* 6:883–93.

89. The invading army is always at a disadvantage in securing water supplies, as proven by the difficulty in supplying adequate water for the British cavalry in its first battle for Gaza on March 26, 1917 C.E., and in previous raids against the Turks in Palestine in September and October 1916 C.E. The horses sometimes went over 30 hours or more without water. Lack of water figured in the British retreat from Gaza and later in Allenby's pursuit of the enemy in southern Palestine in November 1917 C.E. It should be noted, however, that much of the difficulty in obtaining adequate water was due to the destruction of wells and water supply by the Turkish forces who controlled them. Marquess of Anglesey, *A History of the British Cavalry, 1816–1919*, vol. 5: *1914–1919: Egypt, Palestine and Syria* (London: Pen and Sword, 1994) 76, 82, 104, 189.

90. Gal, *Lower Galilee*, 4–7.

Herds of horses could be kept in valleys with caves for protection or easily erected "brush arbors" to protect them from the sun. Horses could be managed for grazing at night by placing hobbles on their feet and using mounted guards, dogs, or donkeys to ward off predators.[91] Mares and foals may also have been secured in dry moats, such as the moat surrounding the main Israelite military base at Jezreel. This dry moat, which was 25 feet wide and 15 feet deep, would have served as an effective subterranean corral.[92]

The Assyrian records indicate that their horse-management procedures included the establishment of logistic bases for the chariots and horses stationed there or passing through on military campaigns.[93] As mentioned above, there is also a detailed memory of a similar infrastructure to support horses in Israel during the Iron Age:

> Solomon had twelve district governors over all Israel, who supplied provisions for the king and the royal household. Each one had to provide the supplies for one month in the year.
>
> Solomon had four thousand stalls for chariot horses, and twelve thousand horses. The district officers, each in his month, supplied provisions for King Solomon and all who came to the king's table. They saw to it that nothing was lacking. *They also brought to the proper place their quotas of barley and straw for the chariot horses and the other horses.* (1 Kgs 4:7, 26–27, NIV)

Baana, son of Ahilud, is listed as the name of the prefect governing Taanach, Megiddo, Jezreel, Beth-shean and Yokneam (1 Kgs 4:12).[94] The strategic location of these particular towns for defense purposes and the suitability of the surrounding terrain for horse-management purposes may indicate a sophisticated, interconnected network of logistic bases for the maintenance and care of large numbers of horses. In Iron Age Israel, the area governed by Baana, now known as the Jezreel Valley, is at least 25 miles square, comprising 16,000 acres of suitable land for raising and caring for horses. In addition, there are references to "chariot cities," presumably established to provide training and stabling for the horses:

91. A single jennet, a female donkey, is frequently used to guard calves in domestic cattle herds against coyotes and other predators. The jennet will chase the invader and attack by kicking and biting, all the while braying loudly and profusely. Her effectiveness is generally unsurpassed.

92. D. Ussishkin and J. Woodhead, "Excavations at Tel Jezreel, 1994–1996: Third Preliminary Report," in *Excavations at Tel Jezreel, 1994–1996: Third Report* (Tel Aviv: Tel Aviv University Institute of Archaeology, 2006) 6–72; I. Finkelstein and N. Silberman, *The Bible Unearthed: Archaeology's New Vision of Ancient Israel and the Origin of Its Sacred Texts* (New York: Free Press, 2001) 186.

93. For a discussion of the logistical challenges encountered by the Assyrian army in gathering barley and straw for feeding their war-horses from the time of Ashurnasirpal II (883–859), see I. Ephᶜal, "Warfare and Military Control in the Ancient Near Eastern Empires: A Research Outline," in *History, Historiography, and Interpretation: Studies in Biblical and Cuneiform Literatures* (ed. H. Tadmor and M. Weinfeld; Jerusalem: Magnes / Leiden: Brill, 1983) 88–106.

94. For a detailed discussion of the Solomonic administrative districts and boundaries, see Rainey and Notley, *Sacred Bridge*, 175–79.

> Solomon assembled chariots and horses. He had 1,400 chariots and 12,000 horses, which he stationed in the chariot towns and with the king in Jerusalem. (1 Kgs 10:26)

Notwithstanding the discrepancy in the number of horses or whether they specifically belonged to Solomon or another king, these verses imply that there was interest in recalling an organized system for caring for horses and maintaining a vibrant chariotry in Iron Age Israel.[95]

An interesting article by archaeologist J. Blakely maps the locations of the tripartite pillared buildings built on major trade routes during the eleventh through ninth centuries.[96] Blakely suggests that these locations, usually identified as commercial entrepots, define the borders of the United Monarchy and emphasize Israel and Judah's continued involvement in protecting international trade routes passing through the region. He posits that these locations may have served as toll-collecting centers and guard stations for the trade routes. He also notes that it is probable that some of these strategically located governmental buildings served as stables for military and defense purposes. Blakely's insights are especially plausible given that providing stabling for horses would not have interfered with any of the other commercial activities that may have occurred simultaneously in these buildings.

Certainly the number of horses in ancient Israel and Judah fluctuated at any given time depending on economic, agricultural, and military realities. However, with the proper infrastructure in place, the land could easily support thousands of horses. Their presence is confirmed by the Assyrian and Aramean inscriptions, Hebrew Bible texts, and the architectural accommodations evident at various archaeological sites in Israel and Judah.

## *Religious Concerns*

It would be easy to assume that the most important function of the horse in ancient Israel was in warfare. However, there is more to the story; horses also play highly significant roles in cultural and religious contexts.[97] In the ancient imagination, beautiful, white horses thundered through the sky, pulling gold and jewel-encrusted chariots, bearing the most precious of all cargo, the very gods of heaven. Meanwhile, on

95. For the suggestion that the list of Solomonic district divisions in 1 Kings 4 closely follows the organization of the Northern Kingdom in the eighth century and may thus better reflect that reality, see I. Finkelstein and N. A. Silberman, *David and Solomon: In Search of the Bible's Sacred Kings and the Roots of the Western Tradition* (New York: Free Press, 2006) 161.

96. J. Blakely, "Reconciling Two Maps: Archaeological Evidence for the Kingdoms of David and Solomon," *BASOR* 327 (2002) 49–54. See also Baruch Halpern, *David's Secret Demons: Messiah, Murder, Traitor, King* (Grand Rapids, MI: Eerdmans, 2001). Holladay, "Stables"; M. Kochavi, "The Eleventh Century B.C.E. Tripartite Pillared Building at Tel Hadar," in *Mediterranean Peoples in Transition* (ed. S. Gitin, A. Mazar, and E. Stern; Jerusalem: Israel Exploration Society, 1998) 468–78.

97. For a review of the horse in ceremonial and religious contexts in the ancient Near East, see U. Seidel, "Pferd. C. Darstellungen," *RlA* 10:490–92.

earth, in efforts to embody the astral vision, priests exercised great care to select only the most excellent horses, befitting inclusion in the sacred mission of transporting the image of the gods in public processions.

In Assyria and Babylonia, pulling the ceremonial chariot bearing the image of Šamaš, Marduk, and Adad was serious and sacred business, requiring lavish preparation. Talented artisans crafted ornate blankets with tassels and intricate harness decorations to caparison the horses formally. The priests conducted complex rituals involving hymns and incantations, some designed to be whispered into the horses' left ears, three times over, while they consumed the special offering set before them:

> You, horse, creature of the holy mountains,
> you are magnificent among all the Pleiades,
> you are assigned in the sky like the rainbow.
> You were born in the holy mountains, you eat pure juniper,
> you drink spring water of the mountains and hills,
> you are given for the chariot of the great lord Marduk,
> you are reckoned with the god for hitching up and unhitching.[98]

These elaborate rituals involving sacred horses are especially interesting when compared with the biblical narrative of King Josiah's purge of the Jerusalem temple:

> He removed from the entrance to the temple of the Lord the horses that the kings of Judah had dedicated to the sun. They were in the court near the room of an official named Nathan-Melech. Josiah then burned the chariots dedicated to the sun. (2 Kgs 23:11, NIV)

Perhaps Judah also had specially revered horses dedicated to pulling the chariot of the Assyrian sun god, thus explaining Josiah's anger and decision to eliminate them. While their memory is preserved, their exact usage remains a mystery.

Likewise, the visions of the postexilic prophet Zechariah include horses as messengers of God on diplomatic missions (Zech 1:8–11) as well as four chariots hitched and ready for use, stationed in the presence of God:

> And I turned, and lifted up mine eyes, and looked, and, behold, there came four chariots out from between two mountains; and the mountains were mountains of brass. In the first chariot were red horses; and in the second chariot black horses; and in the third chariot white horses; and in the fourth chariot grisled and bay horses. Then I answered and said unto the angel that talked with me, "What are these, my lord?" And the angel answered and said unto me, "These are the four spirits of the heavens, which go forth from standing before the Lord of all the earth." (Zech 6:1–5, KJV)

The Psalms also record the idea of a heavenly chariotry:

98. W. G. Lambert, *Babylonian Oracle Questions* (Mesopotamian Civilizations 13; Winona Lake, IN: Eisenbrauns, 2007) 82–83.

> God's chariots are myriads upon myriads, thousands upon thousands;
> the Lord is among them as in Sinai in holiness. (Ps 68:18)

> He makes the clouds his chariot and rides on the wings of the wind. (Ps 104:3, NIV)

The image of God's chariotry is also invoked when Elisha witnesses the "horsemen of Israel" translate Elijah into heaven:

> As they kept on walking and talking, a fiery chariot with fiery horses suddenly appeared and separated one from the other; and Elijah went up to heaven in a whirlwind. Elisha saw it, and he cried out, "Oh, father, father! Israel's chariots and horsemen!" (2 Kgs 2:12)

These heavenly horsemen later reappear in the account of Elisha surrounded by the Aramean chariotry at Dothan:

> When the attendant of the man of God rose early and went outside, he saw a force, with horses and chariots, surrounding the town. "Alas, master, what shall we do?" his servant asked him. "Have no fear," he replied. "There are more on our side than on theirs." Then Elisha prayed: "Lord, open his eyes and let him see." And the Lord opened the servant's eyes and he saw the hills all around Elisha covered with horses and chariots of fire. (2 Kgs 6:15–17)

It is interesting that when the king of Israel visits the prophet Elisha on his deathbed he weeps over him and laments, "My father! My father! The chariots and horsemen of Israel" (2 Kgs 13:14). This epitaph suggests the memory of a connection between the prophet's visions of the heavenly chariotry and the role of the substantive Israelite chariotry led by its earthly ruler. The "horsemen of Israel" occupied more than a military role; they were also associated with divine deliverance and victory.

The concept of heavenly messengers conveyed by horses and chariots developed because the horse was the most powerful and noble domesticated animal, providing speed and invoking awe in the ancient imagination. Driving and riding horses required skill and daring that relatively few brave persons outside a military context ever acquired. Kings and rulers relished their equestrian accomplishments and publicized them broadly. It is not surprising that the horse, which played such an important part in celebrating the majesty of the king and achieving military victories was also assigned a key role in religious thought as an integral and useful component of the divine realm.

## Chapter 4

# *Chariotry in Iron Age Israel*

*The chariots dash about frenzied in the fields,*
*they rush through the meadows.*
*They appear like torches,*
*they race like streaks of lightning.*
*(Nah 2:5)*

The creation of Israel's extensive chariot force required not only a large number of horses but also considerable human capital in the form of charioteers, horse trainers, grooms, stable attendants, and veterinarians.[1] Recognition of the social upheaval inherent in developing an effective state-run chariotry was attributed to the prophet Samuel and used as the primary argument *against* the establishment of a monarchy in Israel:

> This will be the practice of the king who will rule over you: He will take your sons and appoint them as his charioteers and horsemen, and they will serve as outrunners for his chariots. He will appoint them as his chiefs of thousands and of fifties. (1 Sam 8:11–12)[2]

1. Copies of ancient Ugaritic, Akkadian, and Hittite hippiatric texts dealing with medical remedies for sick horses (including enemas of boiled-beer concoctions to cure gastric disease) date to the Bronze and Iron Ages. Earlier examples were excavated in the Assyrian capitals, Aššur and Nineveh. Hittite hippological texts (ca. 1350–1200) dealing with the training of chariot horses have also been found in Hattusa, the Hittite capital. For a detailed description of these various texts, see A. Kammenhuber, "On Hittites, Mitani-Hurrians, Indo-Aryans, and Horse Tablets in the IInd Millennium B.C.," in *Essays on Anatolian Studies in the Second Millennium* (ed. Prince T. Mikasa; Bulletin of the Middle East Culture Center in Japan 3; Wiesbaden: Harrassowitz, 1988) 35–51. See also C. Cohen, "The Ugaritic Hippiatric Texts: Revised Composite Text, Translation and Commentary," *Internationales Jahrbuch für die Altertumskunde Syrien–Palästinas 1996* (UF 28; Münster: Ugarit-Verlag, 1996) 105–54.

2. Human chariot runners were common in Egypt but are not attested in Assyria. Their purpose was to assist in the control of the horses, and may indicate an early stage of inferior horsemanship prior to the advancement of bitting techniques in the Iron Age. See R. Drews, *The End of the Bronze Age* (Princeton: Princeton University Press, 1993) 143. Instead of featuring chariot runners, the Assyrian reliefs typically depict three horses per chariot, two under yoke, and an outrigger hitched to the yoked team by its head stall only; M. A. Littauer and J. H. Crouwel, *Selected Writings on Chariots and Other Early*

However, according to a later biblical text, charioteers and chariotry commanders were considered elite positions during Solomon's reign: "But he did not reduce any Israelites to slavery; they served, rather, as warriors and as his attendants, officials, and officers, and as commanders of his chariotry and cavalry" (1 Kgs 9:22, 2 Chr 8:9). Regardless of whether Solomon or the later Omride Dynasty built the first significant chariotry force in Iron Age Israel, assigning chariotry officers among Israel's elite is in keeping with the military culture of the time.[3] The social prestige and veneration accorded charioteers is also graphically illustrated in Ezek 23:5–27, where political interactions of Jerusalem and Samaria with Assyria and Babylon are compared with illicit sexual relations and described as being inflamed by lust for their handsome chariot officers and mounted horsemen.[4]

Paradoxically, the 1 Samuel text highlights the undesirable burden of raising and maintaining the necessary infrastructure and manpower to support a chariotry corps, while the Kings, Chronicles, and Ezekiel passages emphasize the military hierarchy and high esteem associated with chariotry officers. At a minimum, the social and historical context presupposed by these narratives suggests the memory of an Israelite culture highly engaged with horses and fully acquainted with the special skills required to command them.

---

*Vehicles, Riding and Harness* (ed. P. Raulwing; Culture and History of the Ancient Near East 6: Leiden: Brill, 2002) 253. Three-horse teams were a part of the Ugarit chariotry force during the Late Bronze Age; A. F. Rainey, "The Military Personnel of Ugarit," *JNES* 24 (1965) 17–27.

However, it is not certain that using three horses per chariot was the norm in all cultures. Chariots were also pulled by two horses and accompanied by mounted riders, as depicted in the palace reliefs of Ashurnasirpal (883–859) at Nimrud and of Sargon II (721–705) at Khorsabad; Y. Yadin, *The Art of Warfare in Biblical Lands* (2 vols.; New York: McGraw-Hill, 1963) 2:380–81, 416–17. In theory, the outriggers, when mounted by a cavalryman, would have significantly increased control of the team. The suggestion that the outrigger was a reserve or replacement horse for the horses under yoke is not entirely plausible for two reasons. First, if one of the chariot horses was injured, the charioteer had to drive the chariot from battle and back to camp for assistance in unyoking and reharnessing the reserve horse; such a task was impossible to perform in the heat of battle. Second, the specific horse assigned as the outrigger was probably in that position because he was more suited to it than placement under the yoke. It is more plausible that, if one horse was wounded, the entire team was replaced at the camp where the reserve horses were waiting.

The advantage of three horses pulling a chariot was not increased speed or maneuverability; rather, the benefit was in the extended distance that the horses could cover before tiring; Duncan Noble, "Assyrian Chariotry and Cavalry," *SAAB* 4 (1990) 61-68. To the extent that a battle or skirmish resulted in a hot pursuit, a three-horse chariot could outdistance both two-horse chariots and cavalry, thus increasing the odds of a successful escape. Three horses were also more effective at trampling the enemy underfoot. The disadvantages of three-horse chariots included the increase in space needed to maneuver and the increased difficulty of operation.

3. I. Finkelstein and N. Silberman, *The Bible Unearthed: Archaeology's New Vision of Ancient Israel and the Origin of Its Sacred Texts* (New York: Free Press, 2001) 169–95.

4. Ezekiel's political metaphor for Judah's past relationship with Egypt is even more titillating: "When she was a prostitute in Egypt, there she lusted after her lovers, whose genitals were like those of donkeys and whose emission was like that of horses" (Ezek 23:9, NIV).

## *Origin of Monarchic Chariotry*

It is certain that, during the Bronze Age, chariotry forces were prevalent in the Levant as is evidenced by the Canaanite coalition facing Thutmose III at Megiddo and the various references to chariotry in Israel during the Amarna Age. The chariotry tradition of the Canaanites and Egyptians was well known to the Israelites prior to their development as a nation during the Iron Age. However, emergence of Israel's statehood during the Monarchic period was a necessary precursor to the development of an effective chariotry for military purposes. The care, feeding, and training of horses for military purposes required the sophisticated logistics and coordinated organizational support associated with the special expertise of a centralized government with abundant human resources available for conscription.

Scholars disagree as to whether David, Solomon, or a later king founded the Israelite chariotry, citing lack of archaeological verification and claiming that much of what is written about David and Solomon is by later authors and editors intent on glorifying the early Judahite regency.[5] Exactly which ruler during the Monarchy should be credited with the establishment of Israel's chariotry remains debatable. Nevertheless, in the Hebrew Bible no fewer than 16 kings of Israel and Judah are associated with chariots during the Monarchic period (1025–586), usually in a military context.[6] Regardless of which king is credited, it is certain that an extensive chariotry and supporting infrastructure necessarily existed in Israel and Judah before the Battle of Qarqar in 853.[7] When, where, and how their massive chariotry capabilities

5. Finkelstein and Silberman, *Bible Unearthed,* 144–45; J. B. Burns, "Solomon's Egyptian Horses and Exotic Wives," *Forum 7* (1991) 29–44. See also I. Finkelstein and N. A. Silberman, *David and Solomon: In Search of the Bible's Sacred Kings and the Roots of the Western Tradition* (New York: Free Press, 2006) 154; contra Yadin, *Art of Warfare,* 1:284–86; Y. Ikeda, "Solomon's Trade in Horses and Chariots in Its International Setting," in *Studies in the Period of David and Solomon and Other Essays: Papers Read at the International Symposium for Biblical Studies, Tokyo, 5–7 December, 1979* (ed. T. Ishida; Winona Lake, IN: Eisenbrauns, 1982) 215–38.

6. Saul (1025–1005) fought in chariot battles against the Philistines (1 Sam 13:5, 2 Sam 1:6); David (1005–965) fought in chariot battles against the Arameans (2 Sam 8:4, 10:18); Solomon (968–928) assembled horses, traded them, and established chariot cities (1 Kgs 10:26–29); Rehoboam (928–911) escaped an angry mob in a chariot (1 Kgs 12:18); Asa (908–867) was involved in a chariot battle against the Kushites (2 Chr 14:8–10, 16:8); Zimri (882) served as chariot commander (1 Kgs 16:9); Omri (882–871) commanded the Israelite army, including chariotry (1 Kgs 16:16); Ahab (873–852) died in a chariot battle fighting the Arameans (1 Kgs 22:34–35); Jehosophat (870–846) fought in a chariot battle against the Arameans (1 Kgs 22:29); Jehoram of Judah (851–843) was involved in a chariot battle with the Edomites (2 Kgs 8:21); Ahaziah (843–842), wounded in his chariot, escaped to Megiddo and died there (2 Kgs 9:27); Joram (851–842) was killed by Jehu while driving his chariot (2 Kgs 9:24); Jehu (842–814) conducted a coup from his chariot and trampled Jezebel with his chariot horses (2 Kgs 9:33); Jehoahaz (817–800) lost all but 10 chariots to the Arameans (2 Kgs 13:7); Hezekiah (715-687) lost his chariotry to Sennacherib (2 Kgs 18:23–24); and Josiah (640-609) burned the sun chariots at the temple and died in a chariot battle fighting Egyptian forces (2 Kgs 23:11, 30).

7. For the suggestion that Israel's chariotry infrastructure had to begin development no later than the start of the ninth century, see B. Halpern, "Centre and Sentry: Megiddo's Role in Transit, Administra-

developed is something of a mystery and has received little attention in the scholarly press. However, knowledge of equine physiology, archaeology, and geography provides some provisional clues worth consideration when juxtaposed with hints from the biblical text.

### *Suitability for Chariotry*

Iron Age Israel and Judah were far better suited to accommodate a large chariotry than their surrounding neighbors. They had a varied but compact terrain with open plains for chariot training, rich valleys of pastureland, more than adequate barley and oat production, and hilly areas appropriate for breeding and rearing foals. More importantly, both Israel and Judah were geographically small with short distances between cities, forts, and battlefields. Horses could be moved from one area to another in a matter of *hours*, rather than the weeks required to cross a much larger country such as Assyria or Egypt. Israel and Judah's Iron Age architects exploited these naturally occurring conditions by designing special accommodations for horses and chariotry (for example, stables, training grounds, granaries, and the six-chambered gates discussed later in this chapter).

### *Topography*

The topography of Iron Age Israel, especially in the Galilee with its flat plains and gently sloping hills made chariot travel the simplest and fastest mode of transportation. Horses require neither level roadways nor absolutely flat, pristine plains to pull chariots, and the lightweight battle chariots used by the Israelites did not require a complex system of roads.[8] In fact, overused roadways could become rutted and thus

tion and Trade," in *Megiddo III: The 1992–1996 Seasons* (ed. I. Finkelstein, D. Ussishkin, and B. Halpern; 2 vols.; Jerusalem: Tel Aviv University Press, 2000) 535–77.

8. The Egyptian war chariots weighed only 50–75 lbs.; J. Keegan, *A History of Warfare* (New York: Knopf, 1994) 159. It is reasonable to assume that many Israelite chariots were similar. In fact, the light weight of the chariots was a serious problem; they often bounded out of control and wrecked (Nah 3:2). Consequently, some chariots were fitted with iron or bronze siding (and iron studs on the wheels) to add weight for better traction (Judg 4:3, Nah 2:3). The Assyrians built progressively larger, heavier chariots, sometimes with the capacity to carry four people as illustrated on the palace reliefs of Ashurbanipal; Yadin, *Art of Warfare*, 2:452.

A team of horses can pull an enormous amount of weight. For example, the four-horse team on the night coach between London and Oxford in the late nineteenth century C.E. was pulling about four tons when fully loaded with passengers and luggage; F. T. Underhill, *Driving Horse-Drawn Carriages for Pleasure: The Classic Illustrated Guide to Coaching, Harnessing, Stabling, etc.* (New York: Dover, 1989) 123. An even more significant illustration is the three-horse-team fire-engine brigades prevalent in U.S. cities between 1850 and 1925 C.E. that were required to answer fire alarms at a full-out gallop, pulling six tons of fire-fighting equipment. For a photo essay on this fascinating topic, see C. P. Fox, "Answering the Fire Alarm," in *Working Horses: Looking Back 100 Years to America's Horse Drawn Days* (ed. C. P. Fox; Whitewater, WI: Heart Prairie, 1990) 195–211. By comparison, an Assyrian or Israelite chariot with three armor-clad occupants weighed less than 500 lbs.

more difficult to maneuver than the open plain. The power of the horses could effortlessly carry the charioteer and one or two passengers through shallow streams and up and down sloping hills. The heavily wooded thickets, craggy areas, and marshes could be avoided by the use of an alternative route.[9] The cross-country capability of horses is referenced in Isa 63:13: "Like a horse in open country, they did not stumble."

Chariotry units, especially those traveling en masse, also steered clear of the roadways in an effort to avoid dust. If a unit of chariots traveled down a road in single file, the lead chariot alone would be breathing fresh air and the rest choking on dust. As described in Ezek 26:10, "From the cloud raised by his horses dust shall cover you; from the clatter of horsemen and wheels and chariots, your walls shall shake—." Instead, chariot formations, traveling abreast or fanning out in a v-shape, could be easily accommodated on the vast plains of Megiddo and in the Jezreel Valley. The image of the Israelite chariotry flying across the open plain toward the fortress at Jezreel is captured in the exclamation of the lookout who spotted a fast-moving chariot: "The driving is like that of Jehu son of Nimshi—he drives like a madman" (2 Kgs 9:20).

In fact, speed for ancient chariots was easy; the problem was in stopping them. It appears that the Assyrians developed a combination braking and stabilizing mechanism, depicted in the Lachish reliefs as a long, elliptical, pea-pod-shaped device positioned between the horses on Sennacherib's battle chariot.[10] When deployed upward, this levered device pulled the horses' shoulders together to slow or interrupt their gaits, thus allowing the charioteer to coax them to a stop with voice commands. This "brake," when engaged while the horses were standing, prevented them from taking off until it was lowered.

During the Monarchy, at least in ceremonial contexts, it appears that chariots were sometimes assisted by human "outrunners" alongside the horses, whose function was to clear the way for the horses and also stop them as needed. Both Absalom

9. Competition driving today requires horses to pull coaches and carriages across rough terrain, through water hazards, up steep inclines, and around sharp turns on a rigorous cross-country marathon course of approximately 15 miles to replicate the conditions they faced before modern roadways; E. Peplow, ed., "Carriage Driving," *Encyclopedia of the Horse* (London: Chancellor, 2002) 134–35. The Iron Age chariots were far lighter and more adaptable to the terrain than the wheeled vehicles in use today.

Modern cross-country chariot rides in off-road, all-terrain, lightweight "chariots" pulled by a single Shetland pony climb rocky terrain and claim to achieve speeds of 20 mph; see S. Mullholland, "Saddle Chariots," http://www.saddlechariot.com (accessed 12 October 2007). Furthermore, chariot racing in hippodromes (achieving speeds of 40 mph) is making a resurgence in the Middle East (as well as in France, England, and Brazil); M. Moffett, "The New Ben Hurs: Chariot Racing Stages a Comeback," *Wall Street Journal*, 23 March 2007, p. A1. Reenactments of chariot races in Jerash, Jordan, are sponsored by the Jordan Tourism Board and the Roman Army and Chariot Experience for the entertainment of tourists ("Race: The Roman Army and Chariot Experience," http://www.jerashchariots.com [accessed 22 October 2007]).

10. For illustrations of Assyrian reliefs showing the brake engaged, see Sennacherib Battle Camp on the Lachish reliefs, D. Ussishkin, *The Conquest of Lachish by Sennacherib* (Tel Aviv: Tel Aviv University, Institute of Archaeology, 1982) 77, 116, 118.

and Adonijah are reported to have used as many as 50 chariot runners, ostensibly to enhance their claim to the throne, but also to keep the horses from running away if "spooked" by the crowds (see 2 Sam 15:1, 1 Kgs 1:6). An interesting passage in Jer 12:5 alludes to the job of chariot runner as one that required training: "If you race with the foot-runners and they exhaust you, how then can you compete with horses?" Taken literally, this verse suggests that those chosen to "run with the horses" occupied a position requiring considerable training.

### *Short Distances*

Travel by horsepower was a revolution unequaled in importance until the invention of motorized vehicles in the twentieth century C.E. When Israel's elite shifted from the donkey-riding days of David (2 Sam 16:1–2) to the ceremonial chariotry of Absalom (2 Sam 15:1) and then to the sophisticated war chariots of the Omrides (1 Kgs 16:9), speed of travel increased almost tenfold, reducing travel time by 80 percent over 20 miles or less.[11] The 159-mile (256-km) Via Maris highway, running from Gaza to Dan and linking Egypt with Syria and Mesopotamia, was not just a slow-moving caravan route requiring approximately 5 days of travel when traversing by donkey or camel at 3–4 miles per hour.[12] Rather, chariots, averaging 20 miles per hour, and allowing time for changes of horses, could make the journey in less than a day.[13] With the Israelites' shift to horsepower, the entire distance from northern Israel to southern Judah (Dan to Beer-sheba) could be traveled in 6–7 hours by horseback and less than one day by chariot, assuming changes of horses were available as needed.[14] Chariotry

11. According to the biblical text, David's sons rode mules and donkeys, rather than horses (2 Sam 13:29, 16:2). Absalom's death on a runaway mule indicates the lack of control afforded by inferior bits and perhaps also inferior horsemanship. It is also odd, because typically mules are less prone to spook and run away than horses.

12. Ordinary wayfarers could be expected to average 18 miles per day; L. Casson, *Travel in the Ancient World* (Baltimore: Johns Hopkins University Press, 1994) 53. For calculations of average distances traveled by pack camels under varying weight loads, see J. S. Holladay Jr., "Hezekiah's Tribute, Long-Distance Trade, and the Wealth of Nations ca. 1000-600 B.C.: A New Perspective," in *Confronting the Past: Archaeological and Historical Essays on Ancient Israel in Honor of William G. Dever* (ed. S. Gitin, J. E. Wright, and J. P. Dessel; Winona Lake, IN: Eisenbrauns, 2006) 309–31.

13. Persian express messengers traveled the 1,500 miles of the Royal Road in 9 days with frequent changes of horses and riders (an amazing 184 miles per day), but the normal distance traveled in a day is generally considered to be between 17 and 23 miles; G. Oller, "Messengers and Ambassadors in Ancient Western Asia," *CANE*, 3:1465–73. In rugged terrain, donkey caravans averaged 15–20 miles per day; M. C. Astour, "Overland Trade Routes in Ancient Western Asia," *CANE*, 3:1401–20.

14. The carts pulled by the Chapman horse for traveling salesmen in England in the seventeenth and eighteenth centuries C.E. usually carried loads of 700 lbs. They typically moved at a comfortable pace of "walk five miles in the hour and trot sixteen," thus averaging 20 mph. P. MacGregor-Morris, ed., *The Book of the Horse* (New York: Exeter, 1987) 40. Today, endurance competitions on horseback use one horse and a single rider, usually an Arabian breed, to travel 100 miles over challenging terrain in one day, with stops for mandatory veterinary inspections. E. Peplow, ed., "Endurance Riding," *Encyclopedia of the Horse* (London: Chancellor, 2002) 124–27.

*Table 4.1. Distance and Travel Time from Megiddo to Destinations within Israel*

| *Destinations in Israel* | *Distance* | | *Walking* | | *Donkey* | | *Horse-Drawn Chariot*[a] | | *Rider on Horseback* | |
|---|---|---|---|---|---|---|---|---|---|---|
| *(ordered by ascending distance)* | *(All distances are approx.)* | | *Avg. speed: 3 mi./hr. 4.8 km/hr.* | | *Avg. Speed: 4 mi./hr. 6.4 km/hr.* | | *Avg Speed: 20 mi./hr. 32.2 km/hr.* | | *Avg. Speed: 25 mi./hr. 48.3 km/hr.* | |
| *Jezreel* | 8 mi. | 13 km | 2 hr. | 40 min. | 2 hr. | | | 24 min. | | 19 min. |
| *Beth-Shean* | 19 mi. | 31 km | 6 hr. | 20 min. | 4 hr. | 45 min. | | 57 min. | | 46 min. |
| *Samaria* | 23 mi. | 37 km | 7 hr. | 40 min. | 5 hr. | 45 min. | 1 hr. | 9 min. | | 55 min. |
| *Hazor* | 34 mi. | 55 km | 11 hr. | 20 min. | 8 hr. | 30 min. | 1 hr. | 42 min. | 1 hr. | 22 min. |
| *Ramoth-Gilead (Reimun)* | 43 mi. | 79 km | 14 hr. | 20 min. | 10 hr. | 45 min. | 2 hr. | 9 min. | 1 hr. | 43 min. |
| *Gezer* | 52 mi. | 84 km | 17 hr. | 20 min. | 13 hr. | | 2 hr. | 36 min. | 2 hr. | 5 min. |
| *Dan* | 53 mi. | 85 km | 17 hr. | 40 min. | 13 hr. | 15 min. | 2 hr. | 39 min. | 2 hr. | 7min. |
| *Jerusalem* | 59 mi. | 94 km | 19 hr. | 40 min. | 14 hr. | 45 min. | 2 hr. | 57 min. | 2 hr. | 22 min. |
| *Lachish* | 73 mi. | 117 km | 24 hr. | 20 min. | 18 hr. | 15 min. | 3 hr. | 39 min. | 2 hr. | 55 min. |
| *Gaza* | 86 mi. | 138 km | 28 hr. | 40 min. | 21 hr. | 30 min. | 4 hr. | 18 min. | 3 hr. | 56 min. |
| *Beer-sheba* | 107 mi. | 172 km | 35 hr. | 45 min. | 26 hr. | 45 min. | 5 hr. | 21 min. | 4 hr. | 17 min. |

a. Both chariot and horseback calculations assume that horses could be changed about every hour at stations along the way.

was especially useful in a geographically small country such as Israel, because it improved communication significantly and allowed an effective military defense strategy to be formulated.

As noted in table 4.1, the short distances between towns and fortresses in Israel and Judah made travel by chariot the most feasible means of transportation, at least for the king and military officials. The close locations of the forts in Israel and Judah, strategically placed in distances comfortably achieved in less than an hour by chariot or horseback, meant that it was unnecessary to send written messages, as was often required between the distant Assyrian outposts.[15] A rider could easily remember his message over the short distance and brief time span. Therefore, it is not surprising that the archaeological record in Israel and Judah has yielded no clay-incised "letters" such as are so frequently found in Assyria.

15. B. J. Parker, "Garrisoning the Empire: Aspects of the Construction and Maintenance of Forts on the Assyrian Frontier," *Iraq* 59 (1997) 77–87.

Table 4.1 shows distances from Megiddo, because it was the hub for the main thoroughfares across the southern Levant, centrally located as it was at the crossroads of the major caravan routes, and because it was the largest Iron Age training and stabling facility for chariot horses (see chap. 5).[16] The times given for travel, of course, are based on agreeable weather conditions.

More strategically, the short distances between fortresses allowed Israel and Judah to build and maintain a cohesive standing army. This is significant because it reveals the ease of interplay between Israel's and Judah's armies ("my horses are your horses," 1 Kgs 22:10). Chariots were not only a faster alternative than walking or donkey riding but were also considerably safer and probably more exhilarating than horseback riding. As explained from an engineering perspective: "This means that even at high speeds on irregular surfaces, common not only in Egypt but also in Eastern desert regions, chariots were efficient and comfortable."[17] The ease of travel due to shortened distances also explains the intimate familiarity of the Hebrew prophets with the use of horses in Iron Age Israel: "we shall ride on swift mounts" (Isa 30:16); "the snorting of their horses was heard from Dan" (Jer 8:16); and "they have the appearance of horses; they gallop along like cavalry. With a noise like that of chariots they leap over the mountaintops" (Joel 2:4–5, NIV).[18] When Elijah is said to win a footrace in the wind and rain against Ahab's chariot from Mount Carmel to Jezreel, a distance of more than 18 miles (30 km), we are amazed because he overtook the fastest speed known to humans at that time (1 Kgs 18:44–45).

## *Chariot Cost, Manufacture, and Repair*

In the Bronze Age, chariots were often ornate and costly to build. They also required special craftsmen to maintain them. The expense associated with acquiring and maintaining a chariot is illustrated in one of the Mari texts, where a nobleman, Ila-salim, requests a new chariot from King Zimri-Lim (ca. 1770):

> The king gave me a chariot, but when I went away between the mountains, that chariot broke in the middle, and now as I travel to and fro there is no chariot for me to ride. If it please my lord, may my lord give me another chariot, so that I can organize the country until my lord comes. I am my lord's servant; may my lord not refuse me another chariot.[19]

Of course, all chariots were not created equal. The chariots that were dedicated to the gods and used in religious functions or for state functions were covered in gold

16. A. F. Rainey and R. S. Notley, *The Sacred Bridge* (Jerusalem: Carta, 2006) 176.
17. A. Rovetta, I. Nasry, and A. Helmi, "The Chariots of the Egyptian Pharaoh Tut Ankh Amun in 1337 B.C.: Kinematics and Dynamics," *Mechanism and Machine Theory* 30 (2000) 1013–31.
18. See also Hab 1:8; Nah 3:2–3; Mic 1:13; Zech 1:8–11, 14:20; Ezek 38:15.
19. G. Dossin, *Lettres* (ARM 5; Paris: Geuthner, 1951) no. 66.

and jewels and no doubt required skilled craftsmen to design and build.[20] Ornate chariots were often given as gifts between rulers or as dowries upon marriage.[21] In the epic of Gilgamesh, Ishtar promises Gilgamesh: "I will harness for you a chariot of lapis lazuli and gold, whose wheels are gold and whose horns are brass."[22]

In contrast, ordinary battle chariots were more functional, constructed of simple wood, leather, and linen: just a few sticks of bent wood and leather, with a draft pole and wheels.[23] In the Bronze Age, wheels, axles and other spare parts were carried along on battle campaigns, and inventories of these items were stored in the palaces, as the Linear B tablets indicate.[24] Presumably by the Iron Age, after hundreds of years of practice, an ordinary chariot could be crafted and assembled by anyone with basic carpentry skills. A simple chariot was easy to assemble, dissemble, and transport, yet it was sturdy enough for warfare. On military campaigns in mountainous regions, Tiglath-pileser I (1114–1076) conveyed his chariots with ease:

> Taking my chariotry and army I crossed the Tigris. . . . Riding in my chariot when the way was smooth and going by foot when the way was rough, I passed through the rough terrain of mighty mountains. In Mount Aruma, a difficult area which was impassable for my chariots, I abandoned my chariotry. . . .
>
> Putting my chariotry and army in readiness I took a rugged route between Mount Etnu and Mount Aya. In the high mountains, which cut like the blade of a dagger and which were impassable for my chariots, I put the chariots on (the soldiers') necks (and thereby) passed through the difficult mountain range.[25]

As this passage indicates, chariots were lightweight, serviceable, and durable; they were not complicated machines. It is safe to assume that basic chariots were mass-

20. Another Mari text concerns inscriptions that were to be displayed on the chariot of Nergal, suggesting that the chariots dedicated to the gods were identified in a special manner; J. M. Sasson, "The Lord of Hosts, Seated over the Cherubs," in *Rethinking the Foundations: Historiography in the Ancient World and in the Bible—Essays in Honour of John Van Seters* (ed. S. L. McKenzie and T. Römer; BZAW 294; Berlin: de Gruyter, 2000) 233. See also the elaborate description of Enlil's sacred chariot in Miguel Civil, "Išme-Dagan and Enlil's Chariot," *JAOS* 88 (1968) 3–14.

21. Mitannian King Tushratta sent 10 teams of horses and 10 fully equipped wooden chariots to Amenhotep III (1384–1346) as a wedding present. Likewise, Ashur-uballit, king of Assyria, sent a wedding gift of "a beautiful chariot and two horses" to Akhenaten (1358–1340); T. Bryce, *Letters of the Great Kings of the Ancient Near East: The Royal Correspondence of the Late Bronze Age* (London: Routledge, 2003) 81, 97.

22. A. George, *The Epic of Gilgamesh: A New Translation*, (New York: Barnes & Noble, 1999) VI 9–11, p. 83.

23. B. I. Sandor, "Tutankhamun's Chariots: Secret Treasures of Engineering Mechanics," *Fatigue and Fracture of Engineering Materials and Structures* 27 (2004) 637–46. See also Rovetta, Nasry, and Helmi, "The Chariots of the Egyptian Pharaoh."

24. R. Drews, *The Coming of the Greeks: Indo-European Conquests in the Aegean and the Near East* (Princeton: Princeton University Press, 1988) 84–86.

25. A. K. Grayson, *Assyrian Royal Inscriptions* (2 vols.; Wiesbaden: Harrassowitz, 1972) 1:4 (ii 36); 8, 16 (ii 63); 8, 21 (iii 35).

produced and could be crafted and retrofitted in a week or less from surplus spare parts and inventories.[26] Examples of ordinary battle chariots can be seen on the Balawat gates, the Assyrian reliefs (in the Battle of Lachish, one is tossed down upon the enemy), and the Megiddo ivories.[27]

Evidence on the cost of chariots is sparse: to date, there are only three known textual references to the cost of a chariot during the second and first millennia. An Egyptian text from the thirteenth century places the cost of an ordinary chariot and its pole at 8 *deben* or 64 shekels of silver.[28] A Babylonian boundary stone dating to the reign of Marduk-nadin-ahhe (1098–1081) relates that the price of a chariot and its trappings was 100 shekels of silver.[29]

From Israel, there is reference to a chariot imported from Egypt for 600 shekels of silver (1 Kgs 10:29). Notably, this number is given in the literary context of an ebullient description of Solomon's wealth and fame; the apparent disproportion may emphasize Solomon's extravagance. It may also suggest that Solomon was importing a display chariot, courtesy of Pharaoh, likely decorated with gold and lapis lazuli. It is entirely possible that scholars have been unduly led astray by the exorbitant cost of this single item, the *only* Iron Age reference to the cost of a chariot. This passage could have been written in a much later Babylonian or Persian context, long after the demise of chariot warfare and the widespread use of chariots, when chariots were primarily ceremonial and, therefore, were once again rare and expensive to produce.[30] Significantly, the LXX translation states the cost of Solomon's chariot as 100 shekels rather than 600. Neither price should be interpreted to be a standard price for an ordinary chariot.

There is ample evidence that, throughout the Bronze and Iron Ages, chariot manufacture had become standardized, and chariots were plentiful, easy to repair, and cheap. From the fifteenth-century tomb painting of Hapu (Eighteenth Dynasty), the depiction of chariot-manufacturing shops in Egypt indicates an assembly-line

26. "I will repair his broken wheel within five days, and it will be just as his wheel (used to be)," in E. R. Lacheman, *Administrative Archives (Excavations at Nuzi VI)* (HSS 15; Cambridge: Harvard University Press, 1955) 294:6, 8, 10.

27. Yadin, *Art of Warfare*, 1:242–43. Six chariots were found in the Bronze Age tomb of Tutankhamun (1334–1325), two of which were "state chariots" with gold and colored inlay. It is from these chariots that scholars deduce the weight of ancient chariots. Iron Age battle chariots, however, may have been somewhat more substantial, as can be seen from the Assyrian reliefs. For pictures and discussion of the Egyptian chariots, see R. B. Partridge, *Fighting Pharaohs: Weapons and Warfare in Ancient Egypt* (Manchester, UK: Peartree, 2002) 70–74. For pictures of the slightly more substantial Assyrian chariots, see J. E. Curtis and J. E. Reade, eds., *Art and Empire: Treasures from Assyria in the British Museum* (London: British Museum, 1995) 45, 50–51, 63; and, for Elamite chariots, pp. 73, 76.

28. Ikeda, "Solomon's Trade in Horses and Chariots," 225–26.

29. L. W. King, ed., *Babylonian Boundary-Stones and Memorial-Tablets in the British Museum* (London: British Museum, 1912) 39:15.

30. For the suggestion that the reference to Solomon's chariot was "first and foremost for ceremonial and processional use," see Ikeda, "Solomon's Trade in Horses and Chariots," 224.

production.[31] Linear B tablets found at Knossos, dating to the Mycenean period, show that by the fourteenth century chariot manufacture and repair was standardized, and inventories of spare parts were commonplace. Likewise, Linear B tablets from Pylos, dating to the twelfth century, indicate the prevalence of chariot-wheel manufacture and distribution.[32] The thirteenth-century campsite relief of Ramesses II depicts a mobile chariot workshop repairing a chariot in the midst of the Battle of Qadesh.[33]

According to a letter in Papyrus Anastasi I (late thirteenth century), the chariot-repair center in Joppa (Jaffa) provided all manner of chariot-repair services quickly and efficiently for a hapless Egyptian traveler whose runaway horse smashed his chariot:

> With your chariot lying on its side, you move along swerving to and fro too afraid to pursue your horses. If they are thrown toward the abyss, your horse collar is left exposed and your harness (?) falls. You unharness the team in order to repair the horse collar in the middle of the narrow pass, but you are inexperienced in how to lash it and cannot tie it fast.
>
> You have now entered Joppa and find the meadowland verdant in its season. . . . Your reins have been severed in the darkness, and your team goes off picking up speed (?) over the slippery terrain as the road extends ahead of it. It smashes your chariot and makes [. . .] your leather canteens fall to the ground. . . .
>
> If only you could enter inside the armory with workshops surrounding you and carpenters and leatherworkers in your vicinity, they would do all that you desire. They would take care of your chariot so that it would cease to be inoperative. Your chariot pole would be retrimmed and its supports (?) installed. They would attach leather straps to your horse collar and . . . furnish your yoke. They would mount your chariot case, which has burin engraving [on] the frames. They would attach a pommel to your whip and fasten a lash [to] it. You would then go quickly forth to fight on the battlefield in order to perform heroic deeds.[34]

Whether Joppa remained a chariot-repair center in Israel during the Iron Age is unknown, but the clear implication of the letter is that chariot repair was a relatively quick and simple procedure. Regardless, there is no reason to assume that the craftsmanship for chariot manufacture and repair dissipated until after the demise of chariot warfare in the seventh to fifth centuries.

As previously noted, the cost of Solomon's chariot, 600 shekels (1 Kgs 10:29), is frequently cited for the proposition that Iron Age chariots were expensive to acquire

31. Yadin, *Art of Warfare*, 1:202. For a detailed discussion of the development of Egyptian chariots, yokes, fitting, and harnesses, see A. Herold, "Von Pferdeställen und Wagenteilen," *Achse, Rad, und Wagen* 9 (2001) 4–17.

32. Drews, *The End of the Bronze Age*, 107–9.

33. Yadin, *Art of Warfare*, 1:236–37.

34. "Papyrus Anastasi I: A Satirical Letter," in E. Wente, *Letters from Ancient Egypt* (ed. Edmund S. Meltzer; SBLWAW 1; Atlanta: Scholars Press, 1990) 98–110, esp. p. 108.

and maintain.[35] This may not be the case, at least with regard to Israel. The argument may be made that in the Iron Age not only were chariots easily built and repaired by village craftsmen but also the wood was readily available, the labor was free, and the transportation cost to military headquarters was nil. A chariot was far more valuable to the Assyrians and Egyptians because of their shortage of wood than to the Israelites, who had the raw materials, craftsmanship, and free labor to mass-produce them. During the Bronze and Iron ages, much of the cedar wood for chariot-making came from the forested areas of Lebanon and, therefore, may have been more readily available for the nearby Israelite and Aramean armies.[36] In fact, the wood found near the towns of Taanach and Beth-shean was especially prized for use in manufacturing specific chariot parts.[37]

It has been estimated that maintaining a chariot in battle readiness during the Bronze Age required a half-dozen people, each performing the assigned task of carpenter, harness-maker, metal-worker, groom, or servant.[38] Regardless of their position, these persons were servants to the king; as such, there was no cost for labor, other than perhaps their food. In fact, we may assume that the reason there is no mention of the cost of chariots in the correspondence and literature of the Bronze and Iron ages is because chariots were easily repaired, reproduced, and/or acquired and, thus, not often purchased or sold.[39] The fact that chariots were routinely reused, repaired, and retrofitted strongly lends itself to the presumption that chariots were not shockingly expensive or in short supply.

It is important to remember that far fewer chariots than horses were needed for battle. Notably, chariots, unlike horses, rarely appear in tribute or booty lists as valuable commodities to be especially prized by kings during the Iron Age.[40] Instead,

35. See A. Cotterell, *Chariot: From Chariot to Tank, the Astounding Rise and Fall of the World's First War Machine* (Woodstock, NY: Overlook, 2005) 96–97; Drews, *The End of the Bronze Age*, 110; Nadav Na'aman, "Ahab's Chariot Force at the Battle of Qarqar," in *Ancient Israel and Its Neighbors: Interaction and Counteraction: Collected Essays* (3 vols.; Winona Lake, IN: Eisenbrauns, 2005–6) 1–12. See also Philip J. King and Lawrence E. Stager, *Life in Biblical Israel* (ed. D. A. Knight; Library of Ancient Israel; Louisville: Westminster John Knox, 2001) 244–47.

36. Mahogany used in some Egyptian chariots came from Ethiopia (Sudan); Rovetta, Nasry, and Helmi, "The Chariots of the Egyptian Pharaoh," 1016. See also R. A. Gabriel and K. S. Metz, *From Sumer to Rome* (Westport: Greenwood, 1991) 77.

37. Yadin, *Art of Warfare*, 1:89–90, 236.

38. W. J. Hamblin, *Warfare in the Ancient Near East to 1600 B.C.: Holy Warriors at the Dawn of History* (London: Routledge, 2006) 146.

39. In a similar vein, large pottery craters and storage jugs were universally in use but apparently so easy to reproduce that they were simply made at home, not bought or sold and, therefore, no prices for them exist in the Iron Age.

40. Only Ashurnasirpal II (883–859) claims to have received "a shining chariot" from his expedition to Carchemish and Lebanon as tribute from Sangara, the king of Carchemish. After that, the other Assyrian kings simply claim that they took them in battle. Furthermore, the memory that Joshua burned the enemy chariots after battling the northern kings near Hazor suggests that chariots were not difficult

the Assyrian kings Tiglath-pileser I, Ashurnasirpal II, and Shalmaneser III all claim repeatedly to have captured the chariots and the horses broken to yoke from their enemies. In one battle between Shalmaneser III and Hadadezer of Damascus, the Assyrian king claims to have destroyed the chariots.[41] This probably implies that the Assyrians did not need a surplus of chariots. Furthermore, the mention of thousands of chariots in Egyptian, Aramean, Assyrian, and Hebrew texts (see chap. 6) supports the argument that they were plentiful and readily available and that ownership of chariots was quite commonplace.

## *Chariots in Battle*

The logistical conduct of ancient chariot warfare is a hotly disputed topic.[42] Some authorities suggest that chariots merely transported warriors to combat. Others propose that chariots charged and chased the infantry or that they formed mobile firing platforms from which to fire arrows at the enemy troops. Some scholars limit their role to that of providing a swift exit strategy from the battlefield. There are many opinions, each with its own potential for truth. Although we cannot know the particulars of chariot warfare with certainty in this modern age, we can make some fairly accurate assessments based on the definite limitations of the horses. The war-horse, whether ridden or pulling a chariot, was a lethal weapon, as capable of killing as any arrow or spear, when under the direction of a capable warrior.[43] When two horses were hitched to a chariot carrying soldiers armed with deadly weapons, their killing power was increased exponentially.[44]

to produce when needed (Josh 11:9). Had they been extremely valuable, it would be expected that they would be kept for further use or sale rather than burned.

41. K. Lawson Younger Jr., trans., "Annals: Aššur Clay Tablets" (*COS* 2:113B: 264-66).

42. For a detailed analysis of the way that chariots were used in battle in the Bronze Age and a review of the literature on the subject, see Drews, *The End of the Bronze Age*, 105, 113–29. For a practical discussion arguing against the massed frontal attack and in favor of the mobile firing platform, see A. Hyland, *The Horse in the Ancient World* (Westport, CT: Praeger, 2003) 75–77. For an overview of the use of chariots in Iron Age Assyrian battles, see J. A. Scurlock, "Neo-Assyrian Battle Tactics," in *Crossing Boundaries and Linking Horizons: Studies in Honour of Michael Astour on His 80th Birthday* (ed. G. D. Young, M. W. Chavalas, and R. E. Averbeck; Bethesda, MD: CDL, 1997) 491–597; and Philip Sidnell, *Warhorse: Cavalry in Ancient Warfare* (London: Hambledon Continuum, 2006) 7–17.

43. See chap. 2. A tablet from Tell Leilan in Syria provides textual documentation for the first mounted combat, shortly after the fall of Mari, ca. 1700; see R. Drews, *Early Riders: The Beginnings of Mounted Warfare in Asia and Europe* (New York: Routledge, 2004) 52.

44. Based on radiocarbon dating of Sintashta-culture graves at around 2000, war chariots were first invented in the steppes and introduced to the Near East through Central Asia; D. W. Anthony, *The Horse, the Wheel, and Language: How Bronze-Age Riders from the Eurasian Steppes Shaped the Modern World* (Princeton: Princeton University Press, 2007) 402–3. Anthony argues that the javelin was the weapon of choice for steppe chariot warriors.

The combination of sophisticated, lightweight chariotry and the composite bow revolutionized warfare in the Late Bronze Age.[45] Chariots could go faster and farther than any human or beast and carry two or three people while doing so. They could run down people in seconds and catch people fleeing on donkeys or mules in minutes.[46] Archers in chariots could shoot and kill infantry soldiers from 250 yards and escape unscathed.[47] Chariots could overtake and knock to the ground any infantryman, thereby allowing chariot fighters the advantage in height, angle, and choice of weapon for administering the final blow—all while the infantryman was being held and trampled by the horses' flint-like hooves.

Chariots could also carry many more weapons than could an infantryman or mounted rider; running out of javelins or arrows was not a major concern for chariot fighters. If a chariot did run out of ammunition, the charioteer could promptly race back to the camp and return with more weapons, fresh horses, or fighters. Tactical units of as few as a dozen chariots could neutralize an infantry of thousands on the battlefield, through deadly arrows and ensuing panic.[48] Historian John Keegan suggests that, by encircling the enemy troops with as few as 10 battle chariots, the archers could wound or kill as many as 500 people in 10 minutes.[49] For these reasons, chariot battles prevailed as the preferred warfare convention in open terrain for over 1,000 years.[50]

45. Drews, *The End of the Bronze Age*, 105–6; idem, *Early Riders,* 49. P. R. S. Moorey, "The Emergence of the Light, Horse-Drawn Chariot in the Near East c. 2000–1500 B.C.," in *World Archaeology* 19 (Oct. 1986) 196–215.

46. J. Keegan, *History of Warfare* (New York: Knopf, 1994) 159, 166; S. Weingartner, "Chariots Changed Forever the Way Warfare Was Fought, Strategy Conceived, and Empires Built," *Military Heritage* (August 1999) 19–27.

47. Although some military historians dispute the effectiveness of archers against massed infantry, there is no evidence for an organized phalanx-style infantry in Iron Age Israel; see Gabriel and Metz, *From Sumer to Rome*, 69–72. Oddly, while minimizing the lethal effects of arrow wounds on humans, Gabriel and Metz overemphasize the supposed deadly effect of arrow wounds on horses: "A single arrow impacting upon a horse effectively removed the chariot or cavalryman from the battle" (p. 73). This was not the case. See chap. 2 above for argument to the contrary.

48. "Yet, in every army there is a mob waiting to escape, and its motivation is fear. The real killer on the ancient battle field was fear. Men in combat have their instinctive flight or fight responses held in delicate balance by the thin string of intellect" (Gabriel and Metz, ibid., 84). The motif of panic instigated by divine interference is commonplace in Assyrian and Hebrew lore. For example,

> For the Lord had caused the Aramean camp to hear a sound of chariots, a sound of horses—the din of a huge army. They said to one another, "The king of Israel must have hired the kings of the Hittites and the kings of Mizraim to attack us!" And they fled headlong in the twilight, abandoning their tents and horses and asses—the [entire] camp just as it was—as they fled for their lives. (2 Kgs 7:6–7; see also Jer 4:29, Zech 12:3)

49. Keegan, *History of Warfare*, 166. However, 10 minutes is about how long horses could run at speed in a circle. When they began to slow down, they would become increasingly easy to hit.

50. Chariots were also useful in siege warfare to patrol and block access to a city, as is documented in Hittite accounts (ca. 1650–1600) of the siege of Uršu, where 80 chariots encircled the city; Gary

Hundreds, if not thousands of horses were needed to conduct a chariot battle that lasted more than an hour. Absent divine intervention, the side with the largest number of horses usually won.[51] The importance of a large number of horses in battle strategy is emphasized in Ezekiel:

> For thus said the Lord God: "I will bring from the north, against Tyre, King Nebuchadrezzar of Babylon, a king of kings, *with horses, chariots, and horsemen—a great mass of troops*. . . . From the cloud raised by his horses dust shall cover you; from the chatter of horsemen and wheels and chariots, your walls shall shake—when he enters your gates as men enter a breached city. With the hoofs of his steeds he shall trample all your streets." (Ezek 16:7, 10–11)

However, when considering the dynamics of chariot warfare, one must take into account the endurance levels of the horses and the length of time they could effectively participate in battle. Chariot battles lasting more than one hour were necessarily fought in brief, conjoined segments, with a change of horses occurring approximately every 30 minutes. The duration of each battle depended on the stamina of the horses and, more importantly, on the number of horses available and the ability to deploy chariots effectively and expeditiously under the most arduous conditions.

By the Iron Age, the armies, no doubt, were familiar with the particular conventions and peculiarities specific to chariot warfare, many of which centered on the needs of the horses. For example, to avoid fighting in the heat of the day, the ancients may have commenced battle in the early morning or late evening because the cooler temperature was easier on both humans and animals.[52] Additionally, the practice may have been to convene the armies near each other and then bivouac them for several days before engaging in battle, as reported in the Battle of Aphek (ca. 852): "For seven days [the armies of Ahab and Ben-hadad] camped opposite each other. On the seventh day, the battle was joined" (1 Kgs 20:29). This interval presumably also al-

---

Beckman, "The Siege of Uršu Text (*CTH* 7) and Old Hittite Historiography," *JCS* 47 (1994) 23–34; see also 1 Kgs 20:1, 2 Kgs 6:14. During a siege, it was not safe for anyone to leave on foot because individuals could not outrun the chariot archers and horse patrols. However, some scholars mistakenly assume that chariots had no role in siege operations: "If in pitched battles the cavalry and chariotry have their own role, in the siege and assault of fortresses, the infantry is the sole protagonist of the attacking operations"; D. Nadali, "Assyrians to War: Positions, Patterns and Canons in the Tactics of the Assyrian Armies in the VII Century B.C.," in *Studi in onore di Paolo Matthiae: Presentati in occasione del suo sessantacinquesimo compleanno* (ed. Alessandro Di Ludovico and Davide Nadali; Contributi e materiali di archeologia orientale 10; Rome: Università degli studi di Roma, "La Sapienza," 2005) 167–207.

51. Even the memory of "divine intervention" in the Hebrew Bible text is pictured during the Monarchic period as involving heavenly horses and chariots: "And the Lord opened the servant's eyes and he saw the hills all around Elisha covered with horses and chariots of fire" (2 Kgs 6:7); "God's chariots are myriads upon myriads, thousands upon thousands" (Ps 68:18); "Did you rage against the sea when you rode with your horses and your victorious chariots?" (Hab 3:8, NIV).

52. See Aramean nighttime attack on Dothan (2 Kgs 6:14) and Jehoram's night battle against the Edomites (2 Chr 21:9).

lowed time for negotiations, as well as opportunity for scouting to assess the apparent strength of the enemy, including the number of war-horses available. The relatively short distance from the battlefield to Israel's reserve at the Megiddo or Jezreel horse compound probably made it difficult for the enemy to estimate its true battlefield strength.

We know from the Balawat gates (858–824) that the Assyrian army had both cavalry and chariotry. Each of the battle scenes portrayed on the Balawat gates also shows a nearby campground/bivouac area, either circular or rectangular, and many show horses inside them. Notably, every illustrated pitched battle also had a distant campground/headquarters where the troops ate, slept, and cared for their horses.[53] The Lachish reliefs depicting Sennacherib's battle camps reveal that the horses were sometimes kept in tents.[54] Presumably, these camps were distant enough from the battlefield to discourage the enemy raid but not so far away as to prevent sending reserves and fresh horses to the battlefront as needed. At the Battle of Ramoth-gilead (852), as described in the Hebrew Bible text, the wounded Israelite king commands his charioteer:

> "Turn the horses around and get me outside the camp; I'm wounded." The battle raged all day long, and the king remained propped up in the chariot facing Aram; the blood from the wound ran down into the hollow of the chariot, and at dusk he died. (1 Kgs 22:34–35)

If the battle in fact "raged all day long," numerous trips from the camp to the battlefield for fresh horses were necessary.

It is certain that behind every battle line there was an infrastructure in place to tend to tired horses and to ready new horses.[55] The military encampment portrayed in the Bronze Age Egyptian reliefs at Abu Simbel shows a series of horses engaged in different stages of battle readiness: some next to chariots ready to be hitched, some eating or resting, and still others hitched to chariots engaged in battle. Similarly, the Assyrian palace reliefs and the Balawat gates show active military camps with horses resting and feeding, while other scenes show battles in progress.[56]

The logic in Israel's maintenance of a large chariotry is best explained by Drews: "The rulers must have believed that the chariotry they were so diligently maintaining would in a crisis provide the regime and its subjects with protection and security."[57]

53. See illustration of the "camps" on the Balawat gates in the left-hand corner of pls. 42–43 in Curtis and Reade, eds., *Art and Empire*, 98–99.

54. Yadin, *Art of Warfare*, 2:432; Ussishkin, *The Conquest of Lachish*, 77

55. "Victory normally went to the side which kept in hand a formed reserve in fresh horses after all the opposing cavalry had been committed"; see Adrian Goldsworthy, *Cannae* (London: Cassell Military, 2001) 121.

56. Yadin, *Art of Warfare*, 1:236–37, 2:417.

57. Drews, *The End of the Bronze Age*, 118.

Especially in Israel, with its formidable neighbors able to invade with large, well-trained chariotries, Israel's only possible deterrent was to strive for a superior chariotry. For example, prior to expanding Israel's boundaries during the reign of Jeroboam II (ca. 786), it was necessary to prepare a large chariotry capable of meeting any opposition. This effort required extensive training for horses and men, as well as architectural accommodation.

### *Architectural Advances for the Israelite Chariotry*

Iron Age Israel had at Megiddo the largest stables and most elaborate training complex for chariot horses yet discovered in the ancient Near East.[58] Other architectural advances in use by the Israelite chariotry included four- and six-chambered gates for hitching at city entrances, tripartite pillared buildings for shelter, and granaries constructed at strategic locations throughout Israel during the ninth and eighth centuries.[59] The extent of architectural development dedicated to chariot horses reveals their significance in Iron Age Israel. Intricate and expensive building projects of this sort would be neither necessary nor feasible for an army of only a few hundred horses; Israel's capacity for horse husbandry was virtually unlimited.

### *Six-Chambered Gates*

It was during the late tenth and early ninth century, when chariot warfare was still the conventional means of battle, and travel by chariot was common for the military and the elite that the unique architectural feature of six- (and four-) chambered gates

58. See chap. 5. For the suggestion that the Bronze Age stables of Ramesses II at Pi-Ramses (Qantir) provided shelter for 400–450, see C. Hawley, "Egypt Unearths World's Oldest Stables," *BBC News*, 14 October 1999, http://news.bbc.co.uk/1/hi/world/middle_east/475347.stm (accessed 17 July 2007).

The Iron Age stables excavated at Bastam (modern Iran) accommodated 60 horses. S. Kroll, "Excavations of the Ancient Urartian Fortress Bastam (Iran – Azarbayjan, 1969–1978)," http://www.vaa.fak12.uni-muenchen.de/Iran/Bastam/Bastam.htm (accessed 18 July 2007). The stable complex at Hasanlu (modern Iran), destroyed by fire ca. 800 B.C.E., was large enough to shelter approximately 50 horses. Sargon II's horse compound, Fort Shalmaneser (eighth century), at Nimrud (Kalhu) was used to muster thousands of horses prior to launching campaigns, as was the Review Palace built by Esarhaddon at Nineveh (seventh century); but it is not certain whether they had stables and designated training arenas. S. Dalley and J. N. Postgate, *The Tablets from Fort Shalmaneser* (Cuneiform Texts from Nimrud 3; London: British School of Archaeology in Iraq, 1984) 20.

59. More research about the locations of the granaries in connection with horse facilities is needed. For general discussion, see J. S. Holladay, "Stables, Stables," *ABD* 6:178–83; D. Ussishkin, "Megiddo," *ABD* 4:666–79.

Concerning the tripartite buildings and the debate over their use as stables or storerooms, it is important to consider that they may have been used as both. If the buildings were empty, or partially empty, horses could have been sheltered on one side and the other side used for storage. It is inconceivable that horses would be left to swelter in the hot sun if empty buildings were available for shade. It may be timely to reconsider the usage of tripartite buildings as stables at strategic fortresses with six- and four-chambered gates in Israel and Judah.

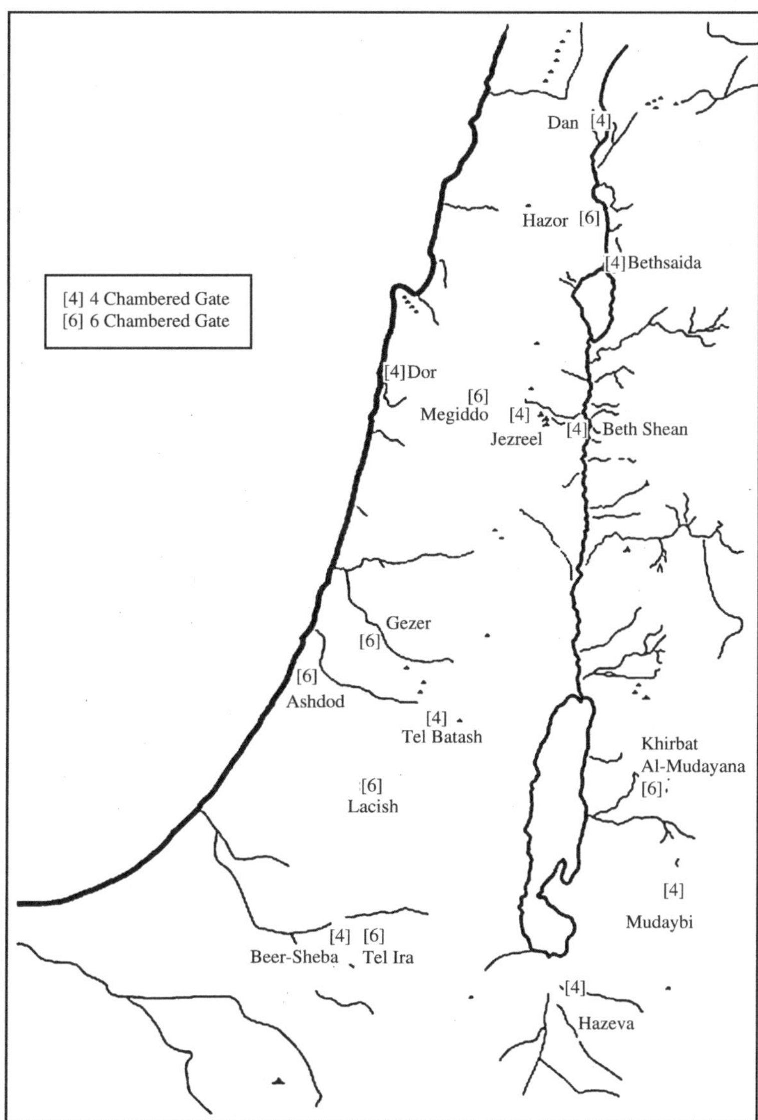


Fig. 4.1. Iron Age locations of 4- and 6-chambered gates in the southern Levant.

began to appear at the entrances of fortresses throughout Israel. These gates were built at Dan, Hazor, Kinrot, Bethsaida, Dor, Megiddo, Jezreel, Beth-shean, Gezer, Ashdod, Tel Batash, Tel Ira, Beer-sheba, and Lachish (fig. 4.1). Similar gates dating to the same time period are found in Moab at Khirbat al-Mudayna and Mudaybi.[60] There are also

60. P. M. Michèle Daviau, "Moab's Northern Border: Khirbat al-Mudayna on the Wadi ath-Thamad," *BA* 60 (1997) 222–28. "Wadi ath-Thamad Project: Jordan, The First Six Seasons," http://www.wlu.ca/page.php?grp_id=296&p=3088 (accessed 18 July 2007). "Mudaybi Gate Complex," http://www.vkrp.org/studies/historical/iron-age-gates/info/mudaybi-gate-complex.asp (accessed 18 July 2007).

chambered gates at the Assyrian forts of Hazeva and Tell el-Kheleifeh (near Aqaba in Jordan), probably built in the late eighth century.[61] Construction of the chambered gates ended in the seventh century (not coincidentally) with the demise of chariot warfare and the rise of cavalry[62] (see fig. 4.1).

Typically, it is suggested that these chambers served as either storage rooms, assembly places for village elders, or secret compartments from which soldiers could ambush enemies.[63] Although these proposals are possible and nonexclusionary, they do not explain the standardization of design of these gates and their placement in strategic locations. Rather, these chambered gates appear to be the Iron Age architects' solution to one of the most vexing challenges of chariotry: how to hitch up numerous chariot horses simultaneously, quickly, and efficiently (see fig. 4.2).

Chariot horses needed to be hitched in pairs when training as military units or when simply traveling from one city to another.[64] In a hot climate such as Israel, chariot horses needed to be hitched, unhitched, and changed at least every 30–60 minutes, depending on how hard and fast the horses were driven. Typically, horses can pull a carriage 20 miles without stopping.[65] The chambered gates in Israel, Judah, and Moab appear to be spaced intentionally to accommodate the physical limitations of chariotry. It is possible that the chambered gates provided an interconnected series of changing stations for chariotry that allowed travel from Beer-sheva to Dan, and all points between, in less than a day's journey (see table 4.2). Each six-chambered gate was conveniently located at an entrance to the fortress, and its design allowed six chariots to be hitched or unhitched in unison with a minimum of personnel.

61. R. Cohen, "Hazeva, Mezad," *NEAEHL* 2:593–94; N. Glueck, "Kheleifeh, Tell El-," *NEAEHL* 3:867-69; G. D. Pratico, "A Reappraisal of Glueck's Conclusions," *NEAEHL* 3:869–70.

62. Iron Age Israel and its warring neighbors were the last stages for chariot battles. Mounted archers began appearing in the Assyrian army as early as the ninth century. By the end of the seventh century, due to improved horsemanship and bitting, chariot warfare was obsolete and was replaced by mounted riders, as discussed in more detail in chap. 7 below. The Median, Persian, Greek, and Roman armies depended on cavalry and infantry, with the heyday of cavalry being from the reign of Sargon II (725–721) to the reign of Darius I (522–486); see Drews, *Early Riders*, 57, 123, 148; S. Dalley, "Ancient Mesopotamian Military Organization," *CANE*, 1:413–22

63. The chambers at Tel Dan are described as "guard rooms" by excavator A. Biran. Additionally, his suggestion that the entrance to Tel Dan was too steep for chariots is mistaken. Horses are easily capable of negotiating inclines while pulling chariots (*Biblical Dan* [Jerusalem: Israel Exploration Society and Hebrew Union College—Jewish Institute of Religion, 1994] 237, 241).

64. From a horse-management perspective, horses typically are not hitched or saddled until they are ready to be used because in hot weather the contact of their skin with the leather harness, yoke, and saddle causes them to sweat, itch, and stomp around, wasting energy. Furthermore, once they are hitched to the chariot and are ready to go, most horses tend to get impatient and jittery. It is not a prudent use of a team's energy to have it driving around in circles, waiting for all the rest of the teams to be hitched.

65. S. Scott, *Plain Buggies: Amish, Mennonite, and Brethren Horse-Drawn Transportation* (Lancaster, PA: Good Books, 1981) 30–31.

*Table 4.2. Approximate Distance (Miles) between Selected Cities with 4*- and 6-Chambered Gates*

| | *Hazor* | *Gezer* | *Megiddo* | *Lachish* | *Ashdod* | *Dan** | *Beth-shean** | *Tel Batash** | *Dor** | *Jezreel** |
|---|---|---|---|---|---|---|---|---|---|---|
| *Hazor* | X | 85 | 34 | 106 | 99 | 19 | 33 | 90 | 45 | 32 |
| *Gezer* | 85 | X | 52 | 21 | 17 | 104 | 56 | 5 | 51 | 54 |
| *Megiddo* | 34 | 52 | X | 73 | 65 | 53 | 19 | 58 | 16 | 8 |
| *Lachish* | 106 | 21 | 73 | X | 17 | 125 | 75 | 15 | 71 | 74 |
| *Ashdod* | 99 | 17 | 65 | 17 | X | 118 | 71 | 15 | 60 | 68 |
| *Dan** | 19 | 104 | 53 | 124 | 118 | X | 52 | 109 | 61 | 50 |
| *Beth-Shean** | 33 | 56 | 19 | 75 | 71 | 52 | X | 60 | 35 | 11 |
| *Tel Batash* | 90 | 5 | 58 | 15 | 15 | 109 | 60 | X | 56 | 59 |
| *Dor** | 45 | 51 | 16 | 71 | 60 | 61 | 35 | 56 | X | 24 |
| *Jezreel** | 32 | 54 | 8 | 74 | 68 | 50 | 11 | 59 | 24 | X |

One of the most difficult aspects of chariotry training was teaching a young horse to stand still while being harnessed to the chariot. In fact, horses must be trained to tolerate and cooperate with the hitching and saddling process.[66] Prior to the construction of the chambered gates, hitching two horses to a chariot in an open space probably required at least six people: two at the heads to bridle the horses and hold them still, two standing at the rear to position the draught pole and keep the horses from swinging out their rear ends (and otherwise interfering with the harnessing process), and two at the withers to attach the harnessing and tighten the yoke. The chariot driver may also have stood in the chariot during this process, creating the proper tension and weight on the yoke while keeping the four long reins straight and out of the way.[67]

The simultaneous hitching of numerous horses in an open area, where they could see each other and feed off of each other's nervousness and misbehavior, would be

66. D. L. Ganton, *Drive On: Training and Showing the Advanced Driving Horse* (Hollywood: Wilshire, 1982) 14–15; D. O. Cantrell and I. Finkelstein, "A Kingdom for a Horse: The Megiddo Stables and Eighth Century Israel," in *Megiddo IV: The 1998–2000 Seasons* (ed. I. Finkelstein, D. Ussishkin, and B. Halpern; 2 vols.; Monograph Series 24; Tel Aviv: Tel Aviv Institute of Archaeology, 2006) 2:643-65.

67. Grooms on an Assyrian relief are depicted struggling to reharness one of Ashurbanipal's chariot horses in the field while on a hunting foray; even with one horse securely yoked, it took three men plus the charioteer to accomplish the task. Note the ratchet-like device in the hands of the grooms used to adjust the tension on the yoke; see J. Reade, *Assyrian Sculpture* (Cambridge: Harvard University Press, 1999) 74; see pl. 83. Various images of chariots and horses appear at http://archaeotype.dalton.org/library/oldsite/seventh.html.

complicated, time-consuming, and fairly dangerous. If a horse became "spooked" or simply decided to be uncooperative, it is likely that other horses would imitate this resistant behavior. This ripple effect could thwart or delay the entire hitching process. It was critical for the yokes to be placed properly and attached securely to avoid the danger of loosening later, causing the draught pole to fall, thence flipping the chariot. It was essential, therefore, that horses be trained or forced to stand quietly while being hitched. The horses' obedience ensured the proper tension in the various harnesses and trappings for the safety of its passengers while underway. In a three-sided chamber, horses were more likely to be cooperative due to the confined space, thereby cutting the hitching time and the requisite number of personnel in half.[68] It is also possible that the more seasoned chariot horses, being well accustomed to the process, were hitched in the three-sided chambers while facing outward toward the aisle. Therefore, the horses diagramed in fig. 4.2 are shown in both poses to illustrate the flexibility that the architecture allowed.

More importantly, from a horse-care perspective, the chambered gates served as a way to process incoming chariotry, enabling horses to be unhitched and cooled down as soon as possible. The cooling-down period after exercise is crucial for hot horses; their muscles must be relaxed slowly while their body temperature falls. Hot horses must not be allowed to stand still because they may "tie-up" with severe stomach cramps that can cause colic and, ultimately, death. For this reason, horses arriving from even a 20-minute canter must be unharnessed quickly, moved out of the way of incoming horses, and hand-walked until they stop sweating. Grooms must also remove salty sweat from the horses by first washing them and/or rubbing them down with straw, and then tending to any skin abrasions that may have developed under the harness. Skin abrasions and sores (sometimes caused by salt accumulation) can attract biting flies, which lay eggs and hatch maggots in the sore, thus seriously afflicting the horses' well-being.

The chambered gates allowed the incoming heated horses in a chariot battalion to be unharnessed expediently with a minimum of personnel and turned over to grooms for care. Although horses might be unhitched anywhere, especially near water or a shaded wood, the chariots and tack also required cleaning and repair after use, because they would be reused on several different teams during a single day. A central hitching/unhitching location was essential. The plaza-like, enclosed compounds near the chambers at Megiddo, Jezreel, Dan, and most other locations provided traffic con-

68. An example illustrates the beauty and practicality of the six-chambered gates. If 32 chariot commanders (1 Kgs 20:1) wanted to travel together from Megiddo to Jezreel, 6 chariots could be hitched and deployed every 10 minutes over the period of an hour. The first group of 6 charioteers would arrive at Jezreel within 30 minutes and begin their unhitching process in the chambers at the Jezreel entrance, followed by the next group 10 minutes later, and so on, with the staggered arrivals over the hour. The entire army command could leave Megiddo and reassemble at Jezreel within 2½ hours and have their horses properly cared for in a timely manner.

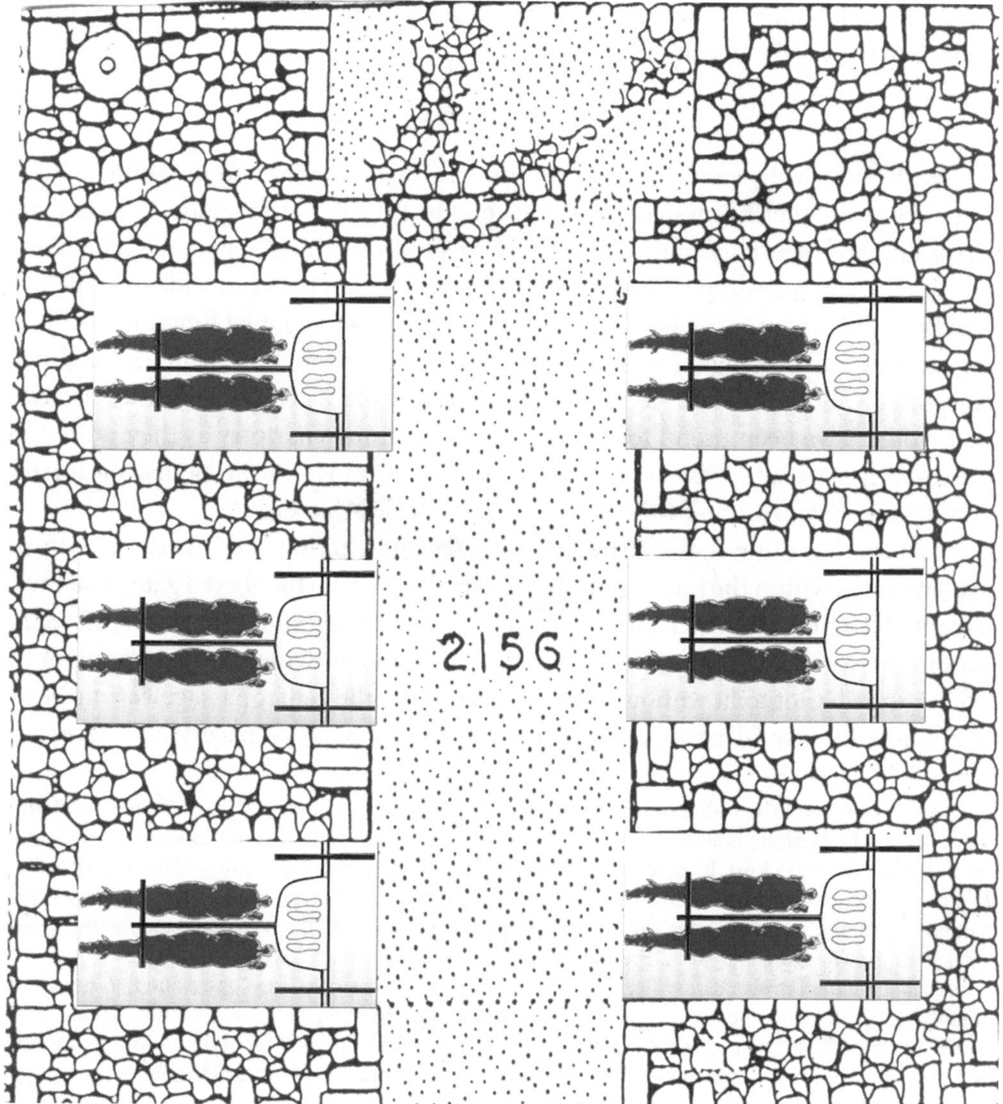


Fig. 4.2. The Megiddo chambered gates with the horses and chariots to scale. Horses are depicted facing both directions because the grooms would decide which direction was most practical for specific horses.

trol and transitory "holding pens" for the horses as they waited to be escorted to their stalls or corral.

From my modern-day visual inspections, the Iron Age locations with the largest, best-designed chariotry-processing entrances, including chambered gates and nearby stables (or buildings that could have been used for stables), include Megiddo, Hazor, Dan, Beth-shean, Bethsaida, Jezreel, Kinrot, Gezer, and Lachish. For horse-processing, these fortress entrances, with adequate room to handle numerous horses at once made excellent sense.

### *Chambered Gate Dimensions*

With their wide center aisle, large enough for incoming and outgoing chariots to pass one another when necessary, the chambered gates at Megiddo, Tel Dan, and Bethsaida are especially impressive.[69] The Megiddo hitching stalls (gate chambers) are 3 m wide by 5 m deep.[70] The extra-deep chambers at Dan (4 m wide by 9 m long) and Bethsaida (3.5 m wide by 10.25 m long) may have served dual functions as hitching stations and overnight corrals.[71] They appear large enough to shelter 10–15 horses comfortably. Both Dan and Bethsaida also had large plaza areas outside the entrances that would have been useful for handling the incoming chariotry as they waited for processing.[72] A water basin found in situ at the entrance to the Bethsaida fortress could have provided water for cooling horses and cleaning tack.[73]

The average dimensions of Iron Age horses and chariots are consistent with the available area within the interior of each of Israel's various chambered gates; even with a team of two horses and a chariot filling a single chamber, sufficient room existed for grooms to work effectively.[74] The length of the chamber provided enough room for the horses and chariot to fit without any protrusion into the center passageway; however, the 5-m-wide central passageway provided additional workspace when necessary.

69. "Gates had to be wide enough to accommodate chariots" (King and Stager, *Life in Biblical Israel,* 236). Bethsaida (Geshur) is strategically located on the north side of the Sea of Galilee, a short distance from Hazor (15 mi., 24 km). It could have served as a mustering point and major granary for a northern coalition army traveling to Ramoth-gilead. According to the chief excavator R. Arav, Bethsaida functioned as an active fortress during the ninth and eighth centuries until the Assyrian conquest by Tiglath-pileser in 732 (R. Arav and R. A. Freund, eds., *Bethsaida: Bethsaida Excavations Project Reports and Contextual Studies* [4 vols.; Kirksville, MO: Thomas Jefferson University Press, 1995–2009] 3:11–15). Arav maintains that the chambers served as granaries and had no military purpose (R. Arav, "Final Report: The City Gate," ibid., 4:1–124, esp. p. 34. However, given the strategic location of Bethsaida in relation to the invading armies of Assyria and Aram, an alternative interpretation is that Bethsaida was an important fortress with accommodation for large numbers of horses for whom the grain would be essential.

70. G. Loud, *Megiddo II: Seasons of 1935–39* (OIP 62; Chicago: University of Chicago Press, 1948) 47.

71. Arav and Freund, *Bethsaida,* 3:27–28, 236–37, pl. 194.

72. Further evidence of an equine presence at Bethsaida is suggested by the one ton of carbonated barley found in chamber three and other indications of grain storage at the gate, the residues of which date to the Iron Age; Arav and Freund, *Bethsaida,* 3:15–16.

73. The Bethsaida water basin is assumed by its excavators to have cultic purposes. However, from my personal inspection in 2006, I believe a plausible and practical alternative suggestion is that the basin held water to wash the faces and feet of incoming horses. The two perforated cups found in the basin would have assisted with this task by allowing the flow of water to wash salty areas on the horses' faces affected by their harnesses. See R. Arav, "Toward a Comprehensive History of Geshur," *Bethsaida,* 1:1–49, especially pp. 19–20.

74. Based on my measurements, a 14.2-hand horse was an average of 2.33 m from nose to tail and 1 m at its widest part.

The basic 5-m chambers could accommodate horses 2.3 m long with attached chariot boxes of 0.50 m (or perhaps 1 m) in depth, with as much as a meter between the rear of the horse and the box (a total of 4.3 m). This spatial layout would still provide grooms with enough area in which to maneuver without halting traffic in the central passageway. The entire gate system seems to have been designed to minimize congestion at the entrance, including the option of "pigeon-holing" as many as six chariot teams at once out of the way of traffic (see fig. 4.2, p. 81).

Likewise, the chambers' wide berth allowed for the accommodation of horses and chariots of varying origin. Specifically, an Egyptian chariot draught pole length (which ran underneath the chariot box) varied from 2.4 to 2.6 m; the axles from 1.92 to 2.36 m. The chariot box was approximately 0.50 m (from back to front) and 1 m wide.[75] Similarly, the wheel-base dimension of the Cypriot chariots, as derived from the eighth-century chariot remains in tombs at Salamis (Cyprus), was 1.54 m.[76] Allowing for an extension of 25 cm of axle length beyond the wheels on each side, the axle length was 2.08 m.[77] Alternatively, the axle length of an eighth-century Assyrian chariot was 1.8 m.[78] A drawing of a Canaanite chariot dating to the early fourteenth century found in an Egyptian tomb depicted an axle length of 1.53 m.[79] Based on the above measurements, the basic 3-m-wide gate chambers could accommodate Egyptian, Cypriot, Assyrian, and Canaanite chariots with space to spare.

The chambered gates at Gezer are especially fascinating. A water trough was found in situ in one of the gates.[80] Hot horses must not be watered immediately, so this large stone basin may have been used to hold water for washing the horses or cleaning their harnesses. Also, an elevated bench running around the inside perimeter on the interior walls of each chamber is evident, which would have allowed grooms the height advantage necessary to securely fasten the yoke over the horse's withers.[81] Why this bench feature was not commonly repeated at other locations is not known; it may be that the stalls with benches running around them were used for hitching the taller horses. Although the average-sized chariot horse is thought to have been 14.2 hands, the large Egyptian (Nubian) horses reached 16 hands.[82] As a practical matter, an average-sized groom (5′6″ to 5′8″) operating at ground level would be

75. For specific dimensions of the eight chariots found in Tutankhamen's tomb, see Rovetta, Nasry, and Helmi, "The Chariots of the Egyptian Pharaoh," 1015, tables 1–2. See also M. A. Littauer and J. H. Crouwel, *Wheeled Vehicles and Ridden Animals in the Ancient Near East* (Leiden: Brill, 1979) 78–80.

76. Ibid., 108.

77. Idem, "Robert Drews and the Role of Chariotry in Bronze Age Greece," in *Selected Writings on Chariots and Other Early Vehicles, Riding and Harness* (ed. P. Raulwing; Culture and History of the Ancient Near East 6; Leiden: Brill, 2002) 66–74.

78. M. E. L. Mallowan, *Nimrud and Its Remains* (3 vols.; New York: Dodd, Mead, 1966) 83.

79. T. Wise, *Ancient Armies of the Middle East* (London: Osprey, 1981) 31.

80. See W. G. Dever et al., "Further Excavations at Gezer, 1967–71," *BA* 34 (1971) 93–132, figs. 1 and 8.

81. Ibid., 114–16.

82. L .A. Heidorn, "The Horses of Kush," *JNES* 56 (1997) 105–14.

*Table 4.3. Iron Age 6- and 4-Chambered Gates (Dimensions Given in Meters)*

| *6-Chambered Gates* | *Passage Width* | *Chamber Width* | *Chamber Depth* |
|---|---|---|---|
| Megiddo IVB | 4.25 | 2.80 | 4.80 |
| Hazor X | 4.20 | 3.00 | 5.00 |
| Gezer | 4.10 | 2.20 | 4.50 |
| Lachish IV | 5.20 | 2.80[a] | 6.00 |
| Ashdod | 4.90 | 3.30[a] | 5.00 |
| Tel Ira | ? | 2.50[a] | ? |
| Tel Batash Area C Strattum III | 4.00 | 2.3–2.7 | 4.70 |
| Mudayna | N/A | 2.0–2.5 | 3.50 |
| *6- or 4-Chambered Gate* | | | |
| Jezreel | 2.5–3.75[c] | 3.12–3.44[c] | 4.38–5.0[c] |
| Quiyafa | 3.90 | 2.50 | 3.50 |
| *4-Chambered Gates* | | | |
| Megiddo IVA | 4.20 | 3.00[b] | 8.20 |
| Beer-sheba V | 4.20 | 3.00 | 6.00 |
| Beer-sheba III | 3.60 | 3.00 | 5.00 |
| Tel Dan | 3.70 | 4.50[b] | 9.00 |
| Ashdod 10 | 4.20 | 2.40 | 3.80 |
| Tell en-Nasbeh (early) | 4.00 | 1.80 | 4.40[b] |
| Dor | 4.50 | 3.70 | 5.80 |
| Tel Batash Area C Stratum | 4.50? | 3.4/4.2 | 4.3/4.3 |
| Hazeva | 4.8–4.0 | 2.50 | 3.30 |
| Kheleifeh | 1.7–3.3 | 2.0–2.5 | 3.3–4.75 |
| Mudaybi | 4.10 | 3.50 | 6.70 |
| Khirbat en-Nahas | 3.63 | 3.00 | 2.90 |

a. Average measurement.
b. Maximum measurement.
c. Measurements from plan.

at a disadvantage in trying to secure a yoke tightly on a horse approaching 16 hands. The bench running around the inside of the chamber at Gezer would have solved this problem by allowing a groom both closer access to the horse's withers and the proper angle for tightening harnesses.

Chambered gates excavated at Ashdod, Gezer, Tel Batash, Tel Ira, Beer-sheba, and Lachish are also conveniently located for horse management and processing. Lachish had an elaborate six-chambered gate located at its entrance and another near its stable and exercise ground during the eighth century. As mentioned by Ussishkin, Lachish may have served as the main horse depot/cavalry center for Jerusalem in the ninth and eighth centuries.[83] This would have made it an especially strategic target for Sennacherib's army during his invasion. More study needs to be done on the chambered gates in Israel, Judah, and Moab to compare their architectural features more precisely, verify their strategic locations, and develop more fully their relationship to the rest of their particular site. It cannot be a coincidence that the uniform dimensions of the chambers at every location perfectly match the measurements of two chariot horses and a chariot, with ample room for the grooms to work.

When considered against the difficulties inherent in hitching horses and the necessity of having several teams leave the compound at once, the chambers provide a creative architectural solution to a potentially serious problem. It is easy to imagine that each location had crews trained to handle horses and process the incoming chariots. Logically, fresh horses were kept waiting at the gate to exchange with incoming ones. This procedure is reflected in Isa 22:7: 'And your choicest valleys were filled with chariots and *the horses were stationed at the gate*' (והפרשים שת שתו השערה; emphasis added). Furthermore, an Assyrian letter dated to the eighth century indicates that the practice at that time was to keep a team of horses hitched up at all times, presumably near the gate, to depart rapidly when needed.[84] Obviously, the ability to hitch and unhitch numerous chariots quickly facilitated the swift and frequent movement of the army from one location to another.

In summary, the six-chambered gates provided six working bays or chambers ideal for handling horses in a safe, secure environment and allowed six chariots to be readied at once. Use of a three-sided chamber as a hitching stall allowed the grooms to align horses evenly, prevent horses from moving away from the yoke during the hitching process, and secure the proper tension on the trappings. Because the horses could see only their teammates and not others being hitched, problems between horses did not materialize. Confinement in a three-sided enclosure had a calming effect on the horses, encouraging them to stand still, and allowing the grooms to work efficiently. Harnessing time was easily cut in half and the needed manpower reduced from six to three people.[85]

83. D. Ussishkin, "The Assyrian Attack on Lachish: The Archaeological Evidence from the Southwest Corner of the Site," *TA* 17 (1990) 84, figs. 2, 14.

84. S. W. Cole and P. Machinist, eds., *Letters from Priests to the Kings Esarhaddon and Assurbanipal* (SAA 13; Helsinki: Helsinki University Press, 1998) no. 88, p. 76.

85. Of course, seasoned, well-trained chariot horses in familiar surroundings could be hitched easily enough without the use of the chambers and no doubt often were. However, the safest, most time-efficient way to hitch them was to use the chambers or perhaps a makeshift three-sided enclosure.

As far as we know, no other nation had as many strategically located chamber-gated entrances as Israel. Their emergence in Iron Age Israel rather than in Egypt or Assyria is probably due to the unique nature of Israel's chariot-friendly topography and the short distances between settlements that made travel by chariot the most expedient mode of transportation. Furthermore, from a military perspective, neither Egypt nor Assyria was under a threat of constant invasion by enemy chariotry forces as Israel was and, therefore, had no particular need for a rapid-deployment defense system. Typically, the Egyptian and Assyrian armies were offensive units that traveled long distances to suitable battlegrounds to stage pitched battles or conduct siege operations with the aim of conquering the lands they invaded.[86] Their horses were led (or ridden) to a camp near the battlefield and on the day of battle hitched and unhitched in the camp before being taken to the field. However, in Iron Age Israel, any fortress or trade caravan was subject to a surprise (relatively speaking) attack. The ability to deploy even six chariots at once was enough to protect the fortress or caravan from an attack by a sizable raiding party.

The abundance of chambered gates in Israel tends to indicate the prevalence of a very large and prominent chariotry. It is also possible that the six-chambered gates were built not only in response to war and threats of war but also to accommodate everyday travel and trade. Regardless, the six-chambered gates are evidence of a substantial horse-management program in Iron Age Israel that, as of yet, remains unequaled elsewhere in antiquity.

86. See chap. 6.

## Chapter 5

# *Stables of Israel: The Case of Megiddo*

*"The capture of Megiddo is the capture of 1,000 cities."*
*(Thutmose III)*[1]

Megiddo is mentioned in at least six separate books of the Hebrew Bible and the New Testament—usually, but not always, in a military context.[2] In addition to Megiddo's historical significance as a battlefield, its strategic location at the crossroads of two great roadways linking north to south and east to west necessarily made it the hub of trade and commerce during the Iron Age. When viewed in the framework of its advantageous location for commerce, the repeated historical battles, and powerful chariotries of its neighbors (Assyria, Egypt, Aram, and Edom), Megiddo logically must have served as one of the pivotal cities in the ancient Near East for supplying and training horses.[3] This conclusion is supported by architectural/archaeological

*Author's Note*: This chapter is based on my contribution "Stable Issues," in the *Megiddo IV* ([ed. I. Finkelstein, D. Ussishkin, and B. Halpern; 2 vols.; Tel Aviv: Tel Aviv University Press, 2006] 2:630–42) archaeological reports and on the chapter that I wrote jointly with I. Finkelstein, "A Kingdom for a Horse: The Megiddo Stables and Eighth Century Israel," in ibid., 2:643–65. I am grateful to Tel Aviv University and to its press for permission to adapt them here.

1. Thutmose III inscribed this sentiment on the walls of the Temple of Amon at Karnak, Egypt, in celebration of his victory against the Canaanites at the Battle of Megiddo in 1479 (J. K. Hoffmeier, trans., "The Annals of Thutmose III," *COS*, 2:2A:7–13).

2. The separate references to Megiddo are as follows: Joshua killed the king of Megiddo during the conquest (Josh 12:7, 12:21); the city of Megiddo was allotted to the tribe of Manasseh (Josh 17:11, 1 Chr 7:29); Deborah and Barak led the Israelites to victory over the Canaanite armies of Sisera by "the waters of Megiddo" (Judg 5:19–20); Solomon made Megiddo one of his district capitals as well as one of his three main fortress cities (1 Kgs 4:12, 9:15); King Josiah of Judah died in battle near Megiddo against Pharaoh Necho in 609 (2 Kgs 23:29–30, 2 Chr 35:20–24); and the eschatological prophecy that military forces will gather in the Valley of Megiddo, at Armageddon (trans. 'hill of Megiddo') before the intervention of Jesus and the armies of heaven riding on white horses (Rev 16:16, 19:14).

3. For a comparison of the Megiddo stables and chariot training facility with later horse compounds in the Roman and Byzantine periods, see Y. Teper, "Stable Facilities in the Land of Israel

evidence: the permanent stabling facilities for as many as 450 horses (150 in the 5 buildings in the Southern Stables, 300 in the 12 buildings in the Northern Stables; see fig. 5.1) and the two large, enclosed, paved courtyards adjacent to them for training and mustering, as well as the stone feeding troughs, the large granary, complex water system, and chambered gates at the entrance.[4]

Initially, the stables found at Megiddo by the University of Chicago archaeologists in the 1920s C.E. were identified as belonging to Solomon, apparently based primarily on the Hebrew Bible (1 Kgs 9:15, 19).[5] However, after further excavations in the 1960s C.E. by Yigael Yadin, the suggestion that the stables more likely dated to the later time of the Omrides and Ahab became generally accepted. However, further analysis by Israel Finkelstein and David Ussishkin of Tel Aviv University led to the conclusion that the palace-like buildings *below* the stables, which were dismantled to build the Northern Stables, should be dated to the ninth-century Omride Dynasty. Therefore, the Northern Stables that replaced these structures had to be later, perhaps around 800, during the reign of Jeroboam II.[6]

However, the story of Megiddo's role as a key equine complex during the Iron Age was temporarily derailed by some scholars in the late twentieth century C.E. James B. Pritchard, a respected scholar but not an expert in equine behavior, launched a controversy over the use of the pillared buildings at Megiddo by questioning their suitability as stables. He expressed concern that the archaeologists had leaped to the "Solomon's-stables" idea because of the biblical text rather than a critical assessment of the evidence. Pritchard raised issues, such as horse removal, tethering hole placement on stone pillars, lack of horse paraphernalia, depth of troughs, plastered and

---

during the Roman–Byzantine Period: Archaeological Data on the Care and Raising of Horses in Israel" (מתקני האורוות בארץ־ישראל בתקופות הרומית והביזנטית המידע הארכיאולוגי על תולדות גידול סוסים בארץ וטיפוחם), in *The Village in Ancient Israel* (ed. S. Dar and Z. Safrai; Tel Aviv: Geographic Research Publications for the Advancement of Knowledge of Eretz Israel, Tel Aviv University, 1997) 230–74.

4. For a discussion of the reevaluation of layout of the Northern Stables and expansive courtyard adjacent, see I. Finkelstein, D. Ussishkin, and B. Halpern, "Archaeological and Historical Conclusions," *Megiddo IV: The 1998–2002 Seasons* (ed. I. Finkelstein, D. Ussishkin, and B. Halpern; 2 vols.; Monograph Series 24; Tel Aviv: Tel Aviv Institute of Archaeology, 2006) 2:843–59. See also Eric Cline, "Area L (The 1998–2000 Season)," in ibid., 1:104–23.

5. P. L. O. Guy, *New Light from Armageddon* (OIC 9; Chicago: University of Chicago Press, 1931) 37–48. R. S. Lamon, and G. M. Shipton, *Megiddo I: Seasons of 1925–34 Stratum I–V* (OIP 42; 2 vols.; Chicago: University of Chicago Press, 1939) 1:32–47, 59.

6. Although it has been suggested that there could be earlier stables under the Stratum IV stables dating to the time of Solomon or Ahab, thus far excavations have not supported the idea; G. I. Davies, "Solomonic Stables at Megiddo After All?" *PEQ* 120 (1988) 130–41. For a detailed discussion of the dating of the Megiddo stables, see Cantrell and Finkelstein, "A Kingdom for a Horse," 644–45. This does not mean that Ahab and the Omrides had no horses at Megiddo; they undoubtedly did keep horses in and around the fortress, albeit without the palatial housing for them that appears later, in the time of Jeroboam II.

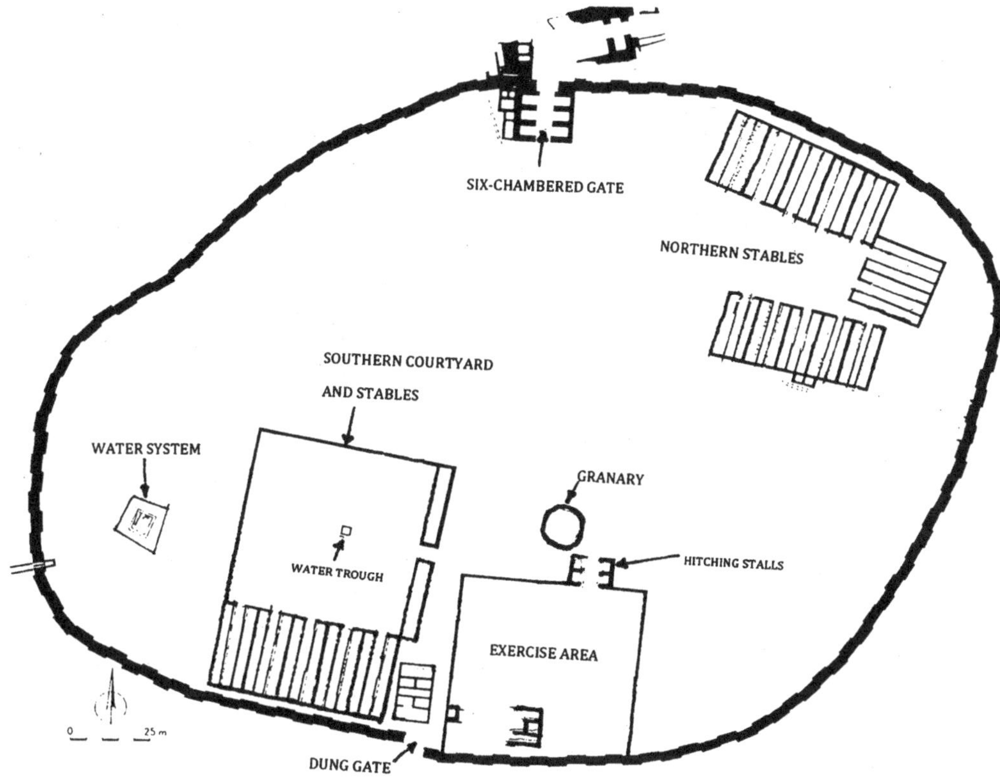


Fig. 5.1. Megiddo stables and training facility, ca. 8th century. Published with the permission of Norma Franklin, The Megiddo Expedition, Tel Aviv University.

paved areas, and suitability of the water tank.[7] Pritchard's article was well received, and other scholars followed with alternative explanations for the tripartite buildings, suggesting their use as barracks, public storehouses, and covered marketplaces.[8]

It became common in academia to refer to the Megiddo stables only with quotation marks around the word "stables" to indicate the widespread belief that, in fact, they were not stables. However, John S. Holladay Jr., a professor and archaeologist from the University of Toronto, who had investigated the Megiddo site several times, compared the structures there to the similar police stables in Toronto. After extensively

7. J. B. Pritchard, "The Megiddo Stables: A Reassessment," in *Near Eastern Archaeology in the Twentieth Century: Essays in Honor of Nelson Glueck* (ed. J. A. Sanders; Garden City, NY: Doubleday, 1970) 268–75.

8. For example, Z. Herzog, "The Storehouses," in *Beer-sheba I: Excavations at Tell Beer-sheba, 1969–1971 Seasons* (ed. Y. Aharoni; Publications of the Institute of Archaeology 2; Tel Aviv: Tel Aviv Institute of Archaeology, 1973) 23–30; Y. Aharoni, *The Archaeology of the Land of Israel: From Historic Beginnings to the First Temple Period* (trans. A. F. Rainey; ed. M. Aharoni; Philadelphia: Westminster, 1982); V. Fritz, "Bestimmung und Herkunft des Pfeilerhauses in Israel," *ZDPV* 93 (1977) 30–40; H. Shanks, "Megiddo Stables or Storehouses?" BAR 2/3 (1976) 12–18.

researching British cavalry manuals, he published a reply to Pritchard's concerns from a horse-management perspective: "The Stables of Ancient Israel: Functional Determinants of Stable Construction and the Interpretation of Pillared Building Remains of the Palestinian Iron Age." This perceptive work answered each of the doubts that Pritchard raised about the Megiddo stable operation, described in detail how a military horse compound operated, and also outlined the evidence for numerous horses and stables in the ancient Near East.[9] As one scholar aptly described Holladay's work:

> He is repeatedly able to show that it was Pritchard's lack of even second-hand knowledge of what goes on in stables which prevented him from appreciating just how closely the pillared buildings at Megiddo and elsewhere match the necessary requirements.[10]

It is puzzling that Holladay's contribution did not put the matter to rest; instead the controversy picked up steam.[11] The idea that the buildings were more suitable as storehouses, public meeting places, or housing for soldiers than stables held fast.[12] Primarily, Pritchard's suggestion that horses could not easily be removed from the stables caused continued confusion, although it was answered precisely in Holladay's work.[13]

At the invitation of Tel Aviv University archaeologists Israel Finkelstein and David Ussishkin, several persons with professional equine experience (riding, training, breeding, and stable management) participated in the 1998 and 2000 seasons at Megiddo (or visited the site during the excavations) and inspected the stable areas.[14] Their observations shed light on the use of these structures as horse-training and breeding centers. These professionals concluded that the Megiddo facilities bore the marks of equine usage, such as crib-biting (discussed below) and pawing.[15] They also

9. J. S. Holladay Jr., "The Stables of Ancient Israel," in *The Archaeology of Jordan and Other Studies Presented to Siegfried H. Horn* (ed. T. L. Geraty and L. G. Herr; Berrien Springs, MI: Andrews University Press, 1986) 103–66.

10. Davies, "Solomonic Stables," 131.

11. For a history of the controversy and analysis of the arguments, see B. Routledge, "For the Sake of Argument: Reflections on the Structure of Argumentation in Syro-Palestinian Archaeology," *PEQ* 127 (1995) 41–49.

12. L. G. Herr, "Tripartite Pillared Buildings and the Market Place in Iron Age Israel," *BASOR* 272 (1988) 47–67; J. D. Currid, "Puzzling Public Buildings," *BAR* 18/1 (1992) 52–61; idem, "Rectangular Storehouse Construction during the Israelite Iron Age," *ZDPV* 108 (1992) 99–121.

13. See also J. S. Holladay, "Stables, Stables," *ABD* 6:178–83; idem, "Stables," *OEANE* 5:69–74.

14. These persons are: Tom Moon (1998), Vanderbilt University, horse breeder; Lord Allenby of Megiddo (2000), an accomplished equestrian and a British cavalry officer; Nancy Later (2000), rider, trainer, international dressage competitor, stable manager; Oded Shimoni (2000), rider, trainer, broker, stable manager and Olympic dressage competitor for Israel; Deborah Cantrell (2000), Vanderbilt University, rider, owner/operator of a horse-training/breeding business, sponsor of an Israeli Olympic dressage team.

15. Basic performance-horse needs have not changed in 3,000 years, and, therefore, comparisons with modern stabling facilities and horse care are relevant. Performance horses, such as racehorses, dressage horses, hunter/jumpers, and driving or "chariot" horses are highly skilled athletes requiring a

noticed special design features, such as the smooth paved center aisle and the stone pillars separating the feeding troughs, consistent with the operation of a stable compound for the training and/or breeding of performance horses.[16] At the 2002 Society of Biblical Literature conference in Toronto, Lord Allenby, Israel Finkelstein, David Ussishkin, Baruch Halpern, Norma Franklin, and I presented findings and conclusions to resolve the long-standing controversy in favor of stables.[17]

## Stabled Horses

In the wild, horses run freely and graze throughout the entire day and a majority of the night. Horses are "trickle feeders," and in their native environment spend an estimated 80% of their time foraging.[18] The stabling of horses alters this natural behavior completely. Stabled horses are confined in a small space and fed limited quantities of concentrated grains and dried grasses only two or three times a day. The rationale behind stabling is not solely confinement; rather, stabling is a key part of a horse's training protocol and psychological conditioning. Stabling teaches wild horses submission and develops trust in human masters. For chariot horses, confinement in a stable with their teammate sharpens the bond between the horses, allowing them to work smoothly side by side.[19]

Although stabling is necessary and beneficial during the training process, confinement of horses is not without its difficulties. Some stabled horses develop psychological problems that are commonly referred to as "stable vices." Stable vices include weaving, box-walking, eating of bedding, eating dung, self-mutilation, wind-sucking, and crib biting or "cribbing."[20]

For the purposes of this discussion, cribbing is the most important of the vices. Cribbing is an action in which a horse grasps a fixed object (such as the edge of a feeding trough or fence post) with its teeth and then arches its neck, thereby forcing open

---

special regimen of exercise and conditioning. Adequate stabling at the training center is a key part of the protocol.

16. For various articles noting their observations, see L. Lexa, "Local Lawyer Searches for Solomon," *The Williamson County Review Appeal*, Williamson Co., TN, 31 July 1998, http://www.reviewappeal.com/073111998/rafamily.html (accessed 12 December 1998); P. J. King and L. E. Stager, *Life in Biblical Israel* (ed. D. A Knight; Library of Ancient Israel; Louisville: Westminster John Knox, 2001) 187–89; D. Cantrell, "Horse Troughs at Megiddo?" *Revelations from Megiddo* 5 (2000) 1–2; H. Shanks, "Horsing Around in Toronto," *BAR* 29/2 (2003) 50–53.

17. Only Halpern expressed reservations about the presence of horses, suggesting instead the possibility of a sheep-raising facility.

18. J. Draper, *The Book of Horses and Horse Care: An Encyclopedia of Horses and a Comprehensive Guide to Horse and Pony Care* (New York: Barnes & Noble, 1998) 200; S. McDonnell, *Understanding Your Horse's Behavior: Your Guide to Horse Healthcare and Management* (Lexington: Eclipse, 1999) 13.

19. *Tactics and Technique of Cavalry* (7th ed.; Harrisburg, PA: Military Science, 1936) 335.

20. Ibid., 320–21.

the soft palate in its throat, to gulp air, usually with a characteristic grunt.[21] Cribbing is typically a repetitive behavior not unlike a child sucking a thumb. Horses are seldom broken of this vice and may crib for hours, entering an almost trance-like state. The signs of a "cribber" in any barn are evident in the damage to stable doors, window ledges, fences, water buckets and especially feeding troughs.[22] It is this cribbing behavior that produced distinctive markings on the inner edges of some of the feeding troughs found at Megiddo and provided the strongest evidence for the protracted stabling of horses there (see fig. 5.2).

### Feeding Troughs

A number of ashlar and mud-brick feeding troughs were discovered at Megiddo. The ashlar troughs were carved from the ashlar building material previously used in the palace buildings. The mud-brick troughs were seen primarily in the Southern Stables. These troughs were each located between stone pillars and were used for allocating grain and hay; any attempt to water from these shallow troughs would have created wet slippery floors in the stalls, posing a danger to the horses.[23] The height (50–65 cm) and width (32–35 cm) of the feeding troughs are adequate to prevent a 14.2-hand horse from jumping out and getting loose.[24]

21. T. Pavord and M. Pavord, *The Complete Equine Veterinary Manual: A Comprehensive and Instant Guide to Equine Health* (Newton Abbot, Devon: David & Charles, 1997) 21.

22. L. Bayley and R. Maxwell, *Understanding Your Horse: How to Overcome Behavior Problems* (North Pomfret, VT: Trafalger Square, 1996) 66–68. The reason for cribbing is to some extent an unsolved mystery. One theory is that the repeated gulping action releases endorphins, giving the horse a mild pleasing sensation that continues as long as it is cribbing; McDonnell, *Understanding Your Horse's Behavior*, 61. Another theory is that the horse is just bored with the confinement and is "playing" a game it cannot quit, perhaps not unlike some children "hooked" on computer video games. Once cribbing is learned, it becomes ingrained behavior for which there is no cure; W. Youatt, *The Horse* (London: Longmans, Green, 1888) 512. Attempts to control cribbing include the use of a muzzle, cribbing strap, and in some severe cases, surgery; J. M. Giffin and T. Gore, *Horse Owner's Veterinary Handbook* (New York: Howell, 1998) 487–89. Nothing has proven totally effective. Fortunately, cribbing does not interfere with riding and training the horse. However, in modern times it is considered a major vice and negatively affects the price of the horse. At the annual horse sales at Keeneland Farms in Kentucky, the auctioneer is required to announce that a particular horse is a "cribber" before the bidding begins (J. J. Sullivan, "Horseman, Pass By," *Harper's Magazine* [2002] 43–59).

23. As noted by Lord Allenby, the horses should not be watered and fed at the same time because negative digestive consequences may occur. The troughs were probably scrubbed out on a regular basis, perhaps with vinegar to dislodge fermenting grain, as was standard procedure in the U. S. Cavalry; War Department, *Cavalry Drill Regulations of United States Army* (New York: D. Appleton, 1916) 370.

24. Although architect Larry Belkin has expressed doubts that the height of the feeding troughs is adequate to prevent the horses from leaving the stall area, this is not a serious concern for several reasons. First, Belkin's reconstruction diagrams incorrectly assume that the horse's feet are standing on bedding several inches above the stone-paved floor rather than on the floor itself. However, the weight of the horse assures that its feet are firmly planted on the stone floor, no matter how deep the bedding is. Therefore, the troughs are at a higher level on the horse than illustrated in his drawings. Second, as discussed in chap. 2, because horses have a blind spot directly in front and under their chin, they cannot see to jump

Fig. 5.2. Evidence of cribbing on a trough from the Southern Stables (photo by the author).

As experienced horse professionals would expect and as mentioned above, a number of the ashlar troughs show evidence of use by a cribbing horse (fig. 5.2). Specifically, in the stable excavated in 1998–2000, in Area L, one trough found in situ shows indentions on the inner lip consistent with cribbing (fig. 5.3).[25] The trough also shows marks of pawing on the front outside surface of the stone, facing the side aisle; these marks are consistent with those made by an impatient horse at feeding time. Some troughs in the southern stable area also show the curved wear and indentions on the inner lips of the feeding troughs that are the certain mark of a cribber (fig. 5.4).[26] In contrast, the troughs that were not used by a cribber retain the original straight, clean line on the edge on the inner lip of the ashlar stone (fig. 5.5).

Pritchard rejected these stone troughs for feeding horses because he believed them to be too shallow (12–14 cm) for that purpose.[27] However, as noted by Holladay, shallow troughs are preferable to deeper food troughs, which allow a horse to dive its mouth deeply into the pile of grain, grab large mouthfuls of feed, and waste the grain by throwing it around and spilling it. Shallow troughs encourage a horse to eat more slowly as it "nuzzles" the grain around the hard surface in order to collect it. In fact, the depth of the stone feeding troughs at Megiddo compares almost exactly with modern-day portable feeding troughs currently sold in tack stores.

---

over the trough. In fact, they would be extremely disinclined to do so because such a leap would be tantamount to jumping down a well. Last, the compact space of the stalls would not provide room for horses to jump forward over the troughs because horses, unlike mules, cannot jump flat-footed; they must have a running start to clear an obstacle. See L. A. Belkin and E. F. Wheeler, "Reconstruction of the Megiddo Stables," in *Megiddo IV: The 1998–2002 Seasons* (ed. I. Finkelstein, D. Ussishkin, and B. Halpern; 2 vols.; Monograph Series 24; Tel Aviv: Tel Aviv Institute of Archaeology, 2006) 2:666–87, figs. 38:2, 38:3, 38:4.

25. The crib marks were independently identified as such, without hesitation by Lord Allenby, Shimoni, Later, and Cantrell.

26. Youatt, *The Horse*, 511–12.

27. Pritchard, "The Megiddo Stables," 271, 275.

Fig. 5.3. Evidence of cribbing and pawing (for the latter, note the arrow) on a trough from Area L (the Northern Stables; photo by the author).

The scene of Ramesses II's battle encampment at Kadesh shows picketed horses eating from troughs (perhaps constructed from mud mortar), attended by watchful grooms with sticks, perhaps to break up food fights.[28] The ninth-century relief of the Assyrian encampment of Ashurnasirpal II from the palace at Nimrud shows horses eating from either side of what appears to be a stone trough similar to troughs found at Megiddo[29] (see fig. 5.6). Although it is unlikely that stone troughs were hauled to the battlefields, it is possible that the artist portrayed a feeding trough typical of the permanent Assyrian stabling facilities. Horses are also shown eating from a ground trough in the Assyrian battle camp ca. 700.[30] A seventh-century Assyrian relief of Ashurbanipal's battle camp also depicts a well-equipped feed cart with hay, grain, and a water jug and shows a feeding trough being filled next to a picketed horse.[31]

The large stone pillars that divide the feeding troughs served another useful horse-management purpose. Because horses exert their hierarchical natures when feeding, the dominant horse feeds first and, if possible, prevents the other horses from eating.[32] However, the square pillars separating the troughs served to block the horses' view of their neighbors while eating, thus preventing biting, snapping, lunging at, and intimidating the horses nearby. The pillars thus averted or reduced the likelihood of injury to neighboring horses during feeding.

28. Horses depicted in the military-camp scene of Ramesses II are feeding from baskets that hang from tripod tethering poles; see Y. Yadin, *The Art of Warfare in Biblical Lands* (New York: McGraw-Hill, 1963) 236. Obviously, horses may be fed in a variety of ways, from leather nose bags to baskets, clay pots, rocks, or simply dumping the grain in a large pile of hay, as is sometimes done at horse shows today.

29. J. Reade, *Assyrian Sculpture* (Cambridge: Harvard University Press, 1999) 41.

30. Andreas Fuchs and Simo Parpola, *The Correspondence of Sargon II, Part III: Letters from Babylonia and the Eastern Provinces* (SAA 25; Helsinki: The Neo-Assyrian Text Corpus Project, 2001) 10, fig. 2.

31. Yadin, *Art of Warfare*, 2:293.

32. A. Hyland, *Equus: The Horse in the Roman World* (New Haven, CT: Yale University Press, 1990) 81.

Fig. 5.4. Evidence of wear on inner lip of a trough from the Southern Stables (photo by the author).

## *Feeding Regimens*

Performance horses must be fed a daily diet rich in protein so they will have the energy to work; they cannot thrive on merely eating grass or hay.[33] Xenophon cautioned Greek cavalry commanders: "While the ranks are filling up, you must see that the horses get enough food to stand hard work, since horses unfit for their work can neither overtake nor escape."[34] The Kikkuli Text, as well as Assyrian, Greek, and Roman sources (and modern practice) confirm the need to stable horses and "grain them up" for several weeks before they are put on active duty.[35] The most practical way to accomplish this is by stabling the horses in order to synchronize the feeding schedules and control the proper mix of rations.

Barley and hay were the primary staples of horse feed throughout antiquity and remain a significant part of the feeding regimen of horses today.[36] The vast, fertile Jezreel Valley near Megiddo could easily supply the barley and hay, in addition to ample

33. Youatt, *The Horse*, 130–38. Conversely, feeding horses grain without hay can cause colic (constipation), which is the number-one cause of death from disease in horses. Because horses cannot vomit, constipation is particularly acute and can lead to a twisted intestine and blockage causing death; M. H. Hayes, *Veterinary Notes for Horse Owners: A Manual of Horse Medicine and Surgery, Written in Popular Language* (rev. J. F. Donald Tutt; 16th ed.; London: Paul, 1968) 393–94.

34. Xenophon, *The Cavalry Commander* 3.235.

35. E. Masson, *L'art de soigner et d'entrainer les chevaux, texte Hittite du maitre écuyer Kikkuli* (Lausanne: Favre, 1998) 43–108; F. M. Fales, "Preparing for War in Assyria," in *Économie antique: La guerre dans les économies antiques* (Entretiens d'archéologie et d'histoire; Saint Bertrand-de-Comminges: Musée archéologique départemental, 2000) 35–62; G. B. Lanfranchi and S. Parpola, *The Correspondence of Sargon II, Part II: Letters from the Northern and Northeastern Provinces* (SAA 5; Helsinki: Neo-Assyrian Text Corpus Project, 1990) no. 3; Xenophon, *The Cavalry Commander* 3.235; Hyland, *Equus*, 43.

36. Barley has a higher energy value than oats and retains its nutritional value longer; Draper, *The Book of Horses*, 200. By contrast, the high gluten content of wheat turns to an indigestible "dough ball" in the horse's stomach and, therefore, may only be fed in small quantities mixed with other feed or diluted with water as bran mash; Giffin, and Gore, *Horse Owner's Veterinary Handbook*, 472.

Fig. 5.5. Clean-cut trough with straight inner edges, not worn by horses on inner lip (photo by the author).

pasturage, required for the healthy maintenance of horses. Ideally, a 1,000-lb. horse eats 12–15 lbs. of grain and 10 lbs. of hay per day.[37] At Megiddo, feeding 24–30 horses per stable × 17 stables twice a day took considerable human effort. The massive granary near the Southern Stables with its capacity of 12,800 bushels would have supplied the 150 horses assigned to this complex for almost a year.[38] However, it is not certain that the horses were stabled year round. The Kikkuli Text suggests that the Hittites employed a fall muster to collect the horses from pasture, followed by a stabling and training period of at least 184 days, presumably to restore them to battle fitness.[39] In Assyria, the widespread muster of horses occurred in the spring before the annual battle campaigns in the summer.[40] It makes sense that also in Israel the horses were stabled in the spring, after the grain harvest, when both barley and wheat (for straw bedding) were readily available.

The barley, grass hay, and bedding straw were likely delivered to the Megiddo stables via donkeys, either by cart or on their backs. Donkeys can carry around 200 lbs. each when evenly balanced, so theoretically one donkey could carry the daily grain ration for an entire stable of 24–30 horses in a single trip.[41] The smooth paved

37. The rations allocated for British cavalry horses in 1917 C.E., for their 70-mile march to attack Beer-sheba included two nose-bags on each saddle, carrying 19 lbs. of grain for two days' provisions. In addition, nearly 100 tons of forage per day were required to feed the 25,000 horses of the Desert Mounted Corps stationed in Palestine; R. M. P. Preston, *The Desert Mounted Corps: An Account of the Cavalry Operations in Syria and Palestine, 1917–1918* (Boston: Houghton Mifflin, 1923) 62. However, the Roman cavalry grain ration was only 3.5 lbs. of barley per day, although the horses were generally smaller than they are today; Karen R. Dixon and Pat Southern, *The Roman Cavalry from the First to the Third Century A.D.* (New York: Routledge, 1992) 210. The barley ration for the Greek cavalry was 6–8 lbs. per day, which was four times the allotment for a soldier; Spence, *Cavalry of Classical Greece*, 281–84.

38. Holladay, "Stables, Stables."

39. Masson, *L'art de soigner*, 108.

40. Fales, "Preparing for War," 39, 43.

41. The homer was derived from the "assload"—the weight that one ass can carry. The "assload," according to M. A. Powell, "lies somewhere in the 90 kg range, fixing the assload of barley at ≈ 150 liters

Fig. 5.6. Water trough in the Southern Courtyard at Megiddo. Published with the permission of the Oriental Institute Museum of the University of Chicago.

center aisle in the Megiddo stables is wide enough for a donkey and small cart to maneuver, and access to the stone troughs is readily accessible from the center aisle. This design remains common in modern stables to simplify the feeding routine.

## *Water Systems*

Watering horses at Megiddo was probably accomplished from two sources: nearby streams and the large sunken cistern (water tank) in the center of the Southern Stables Courtyard[42] (see fig. 5.6). Pritchard argued that the mud-brick construction of the tank made it unsuitable for holding water and suggested that an impermeable material of lime plaster would have been more appropriate.[43] However, horses are notoriously particular about the quality of the water they drink; if they tasted or smelled lime plaster, they would almost certainly refuse to drink. The taste of mud, which is natural to streams and rivers, would encourage rather than deter drinking. Holladay satisfactorily answered Pritchard's argument that the mud brick was not capable of holding water by presenting the example of other mud-brick pools discovered at Amarna that continue to hold water after more than three millennia.[44]

In fact, the large water tank in the Southern Stables Courtyard measured 6.5 ft. (2.3 m) square and 7.5 ft. (2 m) deep. It held 2,775 gallons and, therefore, provided

or the assload of wheat at 120 liters" ("Masse und Gewichte," *RlA* 7:457–517). See also idem, "Weights and Measures," *ABD* 6:897–908.

42. Lamon and Shipton, *Megiddo 1: Seasons of 1925–34 Stratum I–V*, 34, fig 42.

43. Pritchard, "The Megiddo Stables," 272.

44. Holladay, "The Stables of Ancient Israel," 121–22.

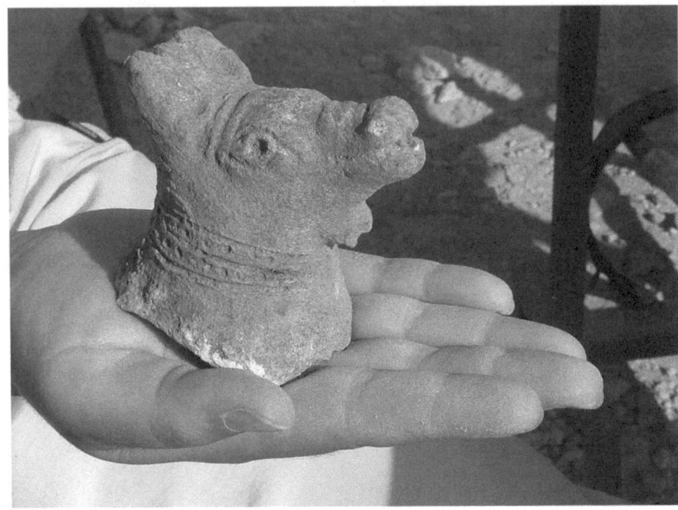

Fig. 5.7. Clay horse head found in stables at Megiddo. Photo by the author.

enough water for nearly 300 horses per day. The size of the water tank allowed several horses to drink at once. It also provided water for bathing and cooling the horses after workouts.[45] Although excavators Lamon and Shipton suggested that the tank was filled once a week from the adjacent water system, it is more likely that it was filled, as needed, throughout the day and covered with a thatched covering to keep it cool during the heat of the day.[46]

Small stones were carefully positioned in the area immediately surrounding the water tank on all four sides to avoid softening the lime plaster surface of the courtyard from water spillage and to give the horses firm footing while drinking. The rub marks that appear on the inner lip of the tank are consistent with marks made by horses when they drink from water troughs as they scratch their chins against the inner edge of the tank.[47]

As a practical matter, the horses were probably led to nearby streams to drink several times a day by grooms, especially if their workouts occurred in the plains outside the fortress.[48] When the horses were trained in the courtyard, using the water tank near the stables was more convenient and undoubtedly saved time and work. It provides further evidence of the sophisticated horse-management program at Megiddo.

45. As a grooming preference, Xenophon suggests washing the horse's head with water rather than brushing, "for it is bony, to clean it with wood or iron would hurt the horse" (Xenophon, *Art of Horsemanship* 5.5, 319). The Kikkuli Text also specifies daily bathing regimens; Masson, *L'art de soigner*, 43–108.

46. Lamon and Shipton, *Megiddo 1: Seasons of 1925–34 Stratum I–V*, 35.

47. These indentations and wear marks can be seen in the photos in excavation reports (ibid.).

48. The longest period a unit of horses went without water in the Palestine campaign in World War I was an astounding 84 hours on the Gaza advance; Marquess of Anglesey, *A History of the British Cavalry*, vol. 5: *1914–1919: Egypt, Palestine and Syria* (London: Pen & Sword, 1994) 188.

## *Tethering Holes*

The pillars between the troughs at Megiddo are no less important than the troughs. In addition to supporting the roof, they served to block the sight of adjacent horses during feeding, thereby preventing the horses from making threatening gestures at one another and fighting for food. At Megiddo, some of the stone pillars between the troughs had tethering holes, most on the sides facing the smooth plastered center aisle.[49] These tethering holes were probably used in the daily grooming process and are consistent with similar holes used in stables today.

It appears logical that the grooms at Megiddo took the horses from the stall areas and groomed them in the center aisle, two or three at a time, single-file in "cross-ties" (tethered ropes attached to either side of the horse's halter) secured to the tethering holes in the stone columns. Grooming in "cross-ties" is a common feature in barns today, because it reduces the manpower involved: no one is required to hold the horse. Grooming in the "cross-ties" is also the safest procedure for the horse, because its head is held up loosely in place, preventing the animal from rubbing its face on its legs, biting at flies, and nipping at the groom. It is likely that the tethering holes also served to secure horses in the center aisle during veterinary inspection and treatment, thereby giving the doctor room to maneuver and placing each horse safely out of the way of others.[50]

In the locations where the tethering holes were pierced on the "stall" side as well as the aisle side, the extra holes may have been used for in-stall grooming or tying an unruly horse. Additionally, if the facility was used as a mare/foal operation, where horses are left at liberty in the open stall area, mares would typically be tied to the troughs during eating to ensure equitable consumption of the grain.[51] The placement of the tethering holes above the troughs is appropriate because its height prevents the horse from pawing and entangling itself in the rope.

The lack of a tethering hole in *each* pillar found at the Megiddo stables served as one of Pritchard's main concerns in his article of refutation.[52] Pritchard noted that the northern stable complex had 54 pillars, of which only 20 were pierced with holes. Pritchard explained the irregular placement of holes in the pillars as an aid in dragging the pillars from the quarry. Notably, Pritchard did not speculate about why the

49. Lamon and Shipton, *Megiddo 1: Seasons of 1925–34 Stratum I–V*, 35.

50. The Ugaritic hippiatric texts are veterinary procedural manuals describing numerous remedies for the treatment of colic (i.e., constipation) and other diseases, which involve treatment by pouring the medicine into the horse's nose, which is similar to modern colic remedies; C. Cohen, "The Ugaritic Hippiatric Texts," *UF* 28 (1996) 110–11.

51. H. Isenbart and E. Bührer, *The Imperial Horse: The Saga of the Lipizzaners* (New York: Knopf, 1986) 60.

52. Pritchard, "The Megiddo Stables," 270, 273.

holes would have been necessary, or even more intriguing, why they were necessary in only *some* of the standardized pillars.

Horse professionals find Pritchard's expectation that *each* pillar in a stable have a tethering hole quite peculiar. The explanation for the irregularity seems simple: not every horse needs to be tied.[53] Each horse is different, with its own varying personalities and peculiarities. Not all horses exhibit the same eating habits. Some have "good manners" and respect their neighbors; others do not. Some horses bite and nip when annoyed; others tolerate almost anything. Because of these individual characteristics, certain horses are tied for particular reasons, such as bad behavior. Additionally, all horses are tied at particular times for very discrete reasons, such as grooming, examining, and minor illness. Notably, not all horses can be groomed or examined at once, no matter how many tethering poles are available.

It seems likely that the tethering holes at Megiddo were bored into the pillars, on site and on an as-needed basis; in this way, the grooms could accommodate the special needs of individual horses and requirements that tethers exist near or in specific stalls. Obviously, the horses in need of tethering could be moved to the locations nearest the holes. We may assume that the number of holes in the pillars were exactly the number needed for the particular complement of horses in that stable; otherwise, they merely would have made another hole where needed, exactly as is done in barns and stables today. It is also possible that in some of the holes a wooden stick was run through to serve as a bridle/halter rack during grooming.

## *Stall Floors*

The paving of the stall (the side aisles) floors in the Northern Stable is of rough stones ranging in diameter from 3 to 5 inches. Xenophon suggested that superior stable construction include paving the stall area with stones to keep horses from slipping and to harden their hooves.[54] He suggested that the stones be "sunk in the floor, touching each other, and about the size of hooves."[55] According to Xenophon, the cobbled-stone stall area would have provided drainage in the absorbent earth between the stones. The paved stall area in the Northern Stables at Megiddo is similar to what was suggested by Xenophon, except that the stones are slightly larger, rough cut, and

53. As can be seen in the interior photograph of the Velbanca at the Lipizzan Stud, a stable housing 10 Lipizzaner breeding stallions, only 2 of the stallions are tied; the others are free in their stalls (Isenbart and Bührer, *The Imperial Horse*, 61).

54. "Now damp and slippery floors ruin even well-formed hoofs. In order that they may not be damp, the floors should have a slope to carry off the wet, and, that they may not be slippery, they should be paved all over with stones, each one about the size of the hoof" (Xenophon, *Art of Horsemanship* 3.315). Modern stable-management treatises also stress the important advice that the stall floor not be slippery; S. Sidney, *The Book of the Horse* (Classic Edition; New York: Bonanza, 1985) 499; H. Wynmalen, *Horse Breeding and Stud Management* (London: Allen, 1971) 43.

55. Xenophon, *Art of Horsemanship* 3.315.

with some space between them. The seventh-century stables in Bastam, Urartu (modern Iran), are similarly constructed, with smooth center aisles and rough-cut stones in the horse standing area.[56]

It is certain that the rough cut of the stones in the stall area, although good for the toughening of the horse's feet, required a covering of deep bedding material such as straw or soil in order to offer the horses a comfortable floor on which to lie without cutting or bruising themselves on the rocks.[57] The straw bedding should have been at least 0.5 m deep. Wheat straw bedding is preferred over oat or barley straw because it is less palatable to horses; thus, the horse is naturally discouraged from eating its bedding.[58]

To prevent disease and reduce the aggravating fly and insect population, the stalls had to be mucked-out daily and the soiled bedding removed.[59] Insect bites cause horses to stomp, kick, bite, and become generally agitated and dissatisfied in their stalls.[60] Typically, horses prefer to urinate on grass or in deeply bedded stalls where the urine is immediately absorbed and does not splash back on their legs.[61] An average horse urinates 5–6 times daily, passing an average of 4.8 liters (8½ pints) of urine every 24 hours. It passes 4–6 manure piles per day, weighing between 44 and 50 lbs. total, or about 8 tons per year.[62] Although not all of the waste would accumulate in the stall, because some would be eliminated while the horse was exercising, the stalls would require round-the-clock maintenance to rid them of manure and urine.[63]

56. S. Kroll, "Bastam: Excavations of the Ancient Urartian Fortress Bastam (Iran – Azarbayjan – 1969–1978)," http://www.vaa.fak12.uni-muenchen.de/Iran/Bastam/Bastam.htm (accessed 17 December 2007).

57. Because horses have the ability to sleep standing up, they only lie down to rest 2–3 hours in a 24-hour period. Horses have a special locking arrangement in the ligaments around their elbow and stifle (knee) joints which prevent them from falling over when asleep. Horses in herds and in stalls take turns lying down to sleep, always leaving a standing horse on guard. Some stalled horses never lie down to sleep; J. Kidd, ed., *The Way of the Horse: How to See the World through His Eyes* (New York: Howell, 1998) 126–27.

58. Draper, *The Book of Horses*, 194–97. It is likely that wheat straw was used at Megiddo because it was readily available from the vast agricultural plains nearby. For ancient stables using dirt as bedding, see Holladay, "Stables," 70.

59. Draper, *The Book of Horses*, 198–99.

60. The elaborate regalia of tassels seen on the chariot horses in the Assyrian reliefs served the functional purpose of fly control. Somewhat similar devices (ear covers with fringed frontlets to keep flies from around the eyes) are used today on competition jumpers.

61. L. Scanlan, *Wild about Horses: Our Timeless Passion for the Horse* (New York: Harper Collins, 1998) 297; M. Eugene Ensminger, *Stockman's Handbook* (Danville, IL: Interstate, 1992) 339.

62. Hayes, *Veterinary Notes*, 19.

63. Generally, modern stables in warm climates are "mucked" or "picked-out" at least three times a day. Herzog's concern that the Megiddo stables did not have drainage in the flanking halls where the horses stood is puzzling for two reasons: (1) most modern stables are not built with drains in the stalls because of the logistical headache of dealing with the inevitability of clogged drains; (2) the danger of injuring the horse if it steps through the drain is not worth the risk. Mucking the stalls on a routine basis

We may speculate that each of the 17 stables had a stable-mucker whose job was to promptly remove waste and soiled bedding and take it from the compound.[64] A small gate near the Southern Stables may have been the designated "dung" gate.[65] Likely the dung, after decomposing on the midden (dung pile), was recycled for fertilizer, fuel, and medicinal purposes. The prompt removal of soiled bedding could explain why higher levels of inorganic residues were not found in the geoarchaeological investigation carried out in the stall areas.[66] Furthermore, the fact that the Megiddo stable floors were never "excavated" as such but were open to the air and elements for over 2,500 years may lessen the expectation of finding an archaeologically significant anthropogenic alteration of the soil.[67]

Pritchard's suggestion that a lime-plastered surface in the stall area would be better suited to horse care is completely inappropriate. In the stables at Megiddo, a lime-plastered surface would have deteriorated due to the effects of urine; and when damp, it would have become slippery and soggy, creating an environment for hoof disease.[68]

---

keeps dampness at a minimum and must be done to remove the manure—for which a drain is useless. Z. Herzog, "Administrative Structures in the Iron Age," in *The Architecture of Ancient Israel: From the Prehistoric to the Persian Periods* (ed. Aharon Kempinski and Ronny Reich; Jerusalem: Israel Exploration Society, 1992) 223–30.

64. For discussion of mucking responsibilities in the Roman army, see Dixon and Southern, *The Roman Cavalry*, 199–200.

65. Lamon and Shipton, *Megiddo 1: Seasons of 1925–34 Stratum I–V*, fig. 34. Dung gates were common features in fortress cities. For references to the Dung Gate in Jerusalem, see Neh 2:13, 3:14.

66. See Cantrell and Finkelstein, "A Kingdom for a Horse," 626. The Iron II stables at Bastam in Urartu show elevated levels of potassium and phosphorus in the stall-area soil; S. Kroll, "Chemische Analysen: Neue Evidenz für Pferdeställe in Urartu und Palästina," *Istanbul Mitteilungen* 39 (1989) 329–33. For a seventh-century horse-training ground adjacent to the Bastam stables, similar to the stable complexes at Megiddo, Jezreel, and Lachish, see Kroll, "Bastam," online. Because mare urine contains significantly more ammonia than that of males, the elevated levels of inorganic residue at Bastam could be attributable to the use of that facility as a mare/foal operation, which also would explain the "double-wide" stall areas; Youatt, *The Horse*, 123; Hayes, *Veterinary Notes*, 19; Holladay, "Stables," 70–72. Furthermore, some of the Bastam stables are not divided by pillars, and the feed troughs are on the outer rather than inner wall, again suggesting a mare/foal operation where all had use of the entire enclosure, except at feeding time, when they were tied, similar to broodmare stables in use today; Isenbart and Bührer, *The Imperial Horse*, 60. See also W. Kleiss, "Bastam, an Urartian Citadel Complex of the Seventh Century B.C.," *AJA* 84 (1980) 299–304. Finally, it should be noted that, in the late eighth century, Urartu was renowned for its government-sponsored horse-breeding program, which according to Sargon II produced foals that developed into superior cavalry horses; H. W. F. Saggs, "Assyrian Warfare in the Sargonid Period," *Iraq* 25 (1963) 145–54.

67. "Most of the anthropogenic elements, compounds, and other chemical characteristics (except soil P) probably do not persist in soils beyond a few thousand years" (V. T. Holliday, *Soils in Archaeological Research* ([Oxford: Oxford University Press, 2004] 304). See also S. F. Cook and R. F. Hetzer, *Studies on the Chemical Analysis of Archaeological Sites* (University of California Publicatiosn in Archaeology 2; Berkeley: University of California Press, 1965) 1–20.

68. Pritchard, "The Megiddo Stables," 272. Because performance horses are often kept stabled 20 hours a day, hoof disease (e.g., thrush, cracked heels)—which in severe cases results in lameness—can become a serious problem if the stall floor is not kept clean and dry of urine and manure; Pavord and Pavord, *Complete Equine*, 102, 200.

In fact, the Megiddo stall areas, with stone embedded in dirt covered in straw bedding would have been the optimum solution for keeping the hooves properly trimmed and clean.[69]

## *Removal of Horses from Stalls*

The dimensions of the Megiddo stalls (2.5–3 m deep) caused the excavators to speculate that, to remove the horse assigned to the back corner, all the other horses would need to be removed first.[70] Although this suggestion gained credence with some commentators, it is not valid.[71] The commentators neglected to consider the natural conformation of a horse. A typical 14.2-hand horse measures 2.33 m from nose to tail but only 1.33 m from chest to tail.[72] Because the area over the troughs was open to the center aisle, a horse could step forward toward the trough until its chest reach the trough, with its neck and head positioned over the trough into the center aisle. This left at least 1.2 m for passage behind the other horses, which are typically less than 1 m wide at their widest part.[73] As noted by Holladay, the horses were probably moved forward or sideways in their stalls to allow room for a horse and groom to pass behind them.[74] With horses accustomed to human handling, this is easily accomplished by a gentle push on the rump or ribs and is standard procedure when someone enters the stall area to muck while the horse is present.[75] Individuals with whom the horses are familiar through their scent and personal contact can move behind them with ease, and there was plenty of room to do so. The idea that the last horses in the aisle were "trapped in" is simply not true.

Horses are instinctively herd animals and naturally form strong bonds to their group.[76] They may also establish an intense preference for one or two other horses,

69. Holladay, "The Stables of Ancient Israel," 134–36.

70. Lamon and Shipton, *Megiddo 1: Seasons of 1925–34 Stratum I–V*, 37.

71. Pritchard, "The Megiddo Stables," 273; K. M. Kenyon, *Royal Cities of the Old Testament* (New York: Schocken, 1971) 97; Herr, "Tripartite Pillared Buildings and the Market Place in Iron Age Palestine"; Z. Herzog, "Administrative Structures in the Iron Age."

72. Based on remains in ritual burial sites, horse sizes in the eighth century ranged from the 10-hand Median horses to the 15-hand Kushite (Egyptian) horses; M. A. Littauer and J. Crouwel, *Wheeled Vehicles and Ridden Animals in the Ancient Near East* (Leiden: Brill, 1979) 111; L. A. Heidorn, "The Horses of Kush," *JNES* 56 (1997) 105–14. It is generally believed that the average horse was 14.2 hands, although variety in size is to be expected. On the whole, they were smaller than today's Olympic horse athletes, which typically vary from 15.5 to over 17 hands.

73. It should be noted that horses are not adverse to small enclosed spaces, as is evident from the ease with which they travel in modern horse trailers with stall widths of no more than 1.5 m.

74. Holladay, *Ancient Stables*, 123. For a graphic depiction (drawn to scale) of the ease with which the horses could be removed from even the smallest stable at Megiddo, see Belkin and Wheeler, "Reconstruction," 673, fig. 38.7.

75. Bayley and Maxwell, *Understanding Your Horse*, 75–76.

76. J. Clutton-Brock, *Horse Power: A History of the Horse and Donkey in Human Societies* (Cambridge: Harvard University Press, 1992) 22.

particularly their stable mates.[77] Removing only one horse from the building as opposed to removing the working unit to which it is accustomed may cause serious disruption in the barn. A perceptive analysis from the fourteenth-century Hittite Kikkuli Text is of special interest on this point: horses worked as teams both in practice exercises with the chariots and during "slow pacing" exercises in which the horses were yoked but driven by a groom on foot.[78] Clearly, the teams were inseparable, and if one horse was a casualty in battle, the survivor would only accept another yokefellow after considerable retraining.[79] For this reason, a "team" generally consisted of three horses: two under yoke and an outrigger, attached by the halter or ridden alongside.[80]

As a practical matter, given the importance of routine in the training of horses, it is most likely that the horses were removed from the stables in established teams of three, beginning with the horses closest to the exit.[81] To the extent that the entire unit was to be trained at once, removing the first three would provide room for grooming the remaining horses in their stalls.[82] For safety and convenience, the actual hitching of horses to chariots logically occurred outside the stable, most likely at specially designated stations such as the six-chambered gates at the entrance to the compound.

### *Presence of "Only" One Exit from the Stable*

The one-exit stable design with a sole opening off the center aisle is viewed by some as being inconsistent with ease in the removal of the horses from the stable.[83] This criticism misses the point from a horse-management perspective. In any horse-training program, it is essential that horses be confined securely and not be allowed out of the stable unless escorted by a groom, for loose horses on a compound are susceptible to injury and may be extremely difficult to catch. Given this necessity, it is logical to conjecture that the stables at Megiddo were specifically designed with only one door. Assuming the one door was secured or guarded, a horse that had escaped from its stall would still have been contained within the stable. The idea was to keep the horses *in*, not to make it easy for them to get *out*.

Of course, the small, narrow doorways, while serving the purpose of securely confining the horses, made their removal problematic in the event of fire. Sadly, nine horses were burned alive at a stable fire in Hasanlu located on the southern shore of Lake Urmia in the Solduz Valley of northwest Iran (ca. 800 B.C.E.). Eight of the horse

77. Oral history is replete with incidents of runaway horses that return to the barn to reunite with their companions, carrying the hapless rider against his/her will.

78. Masson, *L'art de soigner*, 43–108.

79. A. A. Dent, *The Horse through Fifty Centuries of Civilization* (New York: Holt, Rinehart, and Winston, 1974) 59.

80. Littauer and Crouwel, *Wheeled Vehicles and Ridden Animals*, 114.

81. Kenyon, *Royal Cities*, 97.

82. The smallest tactical unit was a group of 10 chariots; an Egyptian squadron was made up of 50 chariots; R. Drews, *The End of the Bronze Age* (Princeton: Princeton University Press, 1993) 126.

83. Herzog, "Administrative Structures in the Iron Age," 226–27.

skeletons were found in two stables with extremely narrow doors and an additional maze-like entrance feature designed to further ensure their containment. The skeleton of a man carrying two bridles, presumably on his way to rescuing the trapped horses was excavated in the stable corridors.[84] Given the size of the Hasanlu stables, we may conclude that the majority of the horses were evacuated successfully or were absent when the fire occurred. It is generally thought that the conflagration, which also resulted in at least 246 human deaths, was intentionally set by an invading army from Urartu.[85] Fortunately, there is no evidence that a fire ever occurred near the Megiddo stables.

## *Stabling of Stallions*

Because many historians assume that all chariot horses were stallions, to the visiting horse professionals the most intriguing question about the Megiddo stables concerned the practical problems inherent in stabling stallions. Horse professionals are divided in opinion on whether stallions, mares, and geldings can be kept peacefully in the same stable. One respected authority, Ann Hyland, states that she has kept "stallions, geldings and mares in close proximity for decades and never had any problems."[86] However, another modern expert notes that "stallions used for racing or showing are often very aggressive and difficult to manage when around other stallions or mares."[87] By comparison, mares and male horses were stabled together in the Roman army, apparently without difficulty.[88]

Although stallions are highly prized for their intelligence and stamina and often make brilliant show horses, their sexual urges can make them a safety risk.[89] In modern barns, it is customary to take special precautions in the stabling of stallions because, although many are well behaved around people, they can become unruly when aroused by a mare in season.[90] Thus, one may ask what prevented the stallions stabled at Megiddo from fighting in their stalls?

Modern stables typically have permanent stall dividers between the horses that prevent them from touching each other.[91] It appears that the Megiddo stables did not

84. Maude de Schauensee, "Horse Gear from Hasanlu," *Expedition* 31/2–3 (1989) 37–52.

85. Oscar White Muscarella, "Warfare at Hasanlu in the Late 9th Century," *Expedition* 31/2–3 (1989) 24–36.

86. Hyland, *Equus*, 81.

87. E. Squires, *Understanding the Stallion* (Lexington, KY: Eclipse, 1999) 75.

88. Dixon and Southern, *The Roman Cavalry*, 177.

89. For other stallion vices (biting, rearing, charging, masturbation, etc.), see Giffin and Gore, *Horse Owner's Veterinary Handbook*, 364–66. In the wild, stallions sometimes rape mares and kill foals during harem takeovers (Scanlan, *Wild About Horses*, 299).

90. D. Morris, *Horsewatching* (New York: Crown, 1988) 35–36.

91. Only a few stables, such as the Spanish Riding School in Vienna, home of the famous Lipizzaner stallions, use all stallions as performance horses (Isenbart and Bührer, *The Imperial Horse*, 200). Clay Harlin, a respected horse-breeder in Franklin, Tennessee, reports that on a trip to Yugoslavia in 2000 C.E. he visited stallion training stables with no partitions between the horses. When he questioned whether

have permanent stall dividers although, as suggested by Holladay, they probably had some type of portable "bail" separators hanging by ropes from the ceiling.[92] Even so, it seems that stallions stationed side by side in proximity without dividers must have been from a breed that was calm by nature and were highly trained, with good "ground manners," and were watched by attentive grooms. It is likely that the ancient stable masters used "tricks of the trade" to control the sexual urges of the stallions, such as equipping them with stud guards and/or rubbing pungent ointments in their nostrils to interfere with the scent of the mares.[93]

At Megiddo, there was ample room for both a training operation, feasibly in the Southern Stables' courtyard, and a breeding operation, perhaps in the Northern Stables. The Northern Stables also had a smaller smooth, paved courtyard that could have been used as a paddock for turnout.[94] Certainly the horses could have been separated by sex if necessary in the 17 barns available.

### *Lack of Horse Paraphernalia at Megiddo*

Pritchard noted the absence of bits, blinkers, and decorative horse trappings as proof that the Megiddo pillared buildings were not used as stables. When one considers the ornate fittings depicted on the reliefs of Assyrian cavalry of the ninth century, it is reasonable to expect that remnants of these trappings would be found in an area given over to the housing of horses.[95] However, Pritchard apparently did not consider the idea that the Assyrian reliefs also had both decorative and propagandistic aims. The horses in full dress regalia seen on the Assyrian reliefs displayed the might of Assyria in martial mode rather than portraying a realistic scene.[96] Elaborately decorated harnesses and accoutrements should not be expected at a training center where the emphasis is more on practicalities than ornamentation.

In fact, only three Iron Age sites have yielded extensive horse trappings and paraphernalia: the tombs at El Kurru of horses belonging to Kushite kings (discussed in chap. 3 above), the eighth- and seventh-century horse burials associated with prominent people on Cyprus, and the Hasanlu fire that burned horses, stables, and the ad-

---

there was ever conflict between the horses, he was told that trouble only occurred when a new groom forgot the precise order of the horses and accidentally put one next to a stranger (private communication).

92. Holladay, "Stables," 138, fig. 3.

93. Both measures are used today. Stud guards (leather devices with protruding nails that may be strapped to the stallion's stomach to prevent erections) may be purchased at tack stores.

94. Lamon and Shipton, *Megiddo 1: Seasons of 1925–34 Stratum I–V*, 47. A sounding undertaken in Area L in 2004 has proven that Building 434, located in the space between the three sets of northern stables, belongs to Stratum III (see chaps. 8 and 43). In the time of Stratum IVA, this entire space served as a large open, pebble-paved courtyard that sloped downhill all the way to the city gate.

95. Pritchard, "The Megiddo Stables," 273.

96. J. E. Reade, "The Neo-Assyrian Court and Army: Evidence from the Sculptures," *Iraq* 34 (1972) 87–112.

joining tack room (discussed above).[97] In the case of the horse burials, the trappings and ornamentation were intentionally left on the horses as symbols of the owners' wealth and power.[98] In the Hasanlu fire, which also killed many human inhabitants, there was no time to evacuate the horses or the tack stored there. These circumstances did not apply to Megiddo, where the approach of the Assyrian army was anticipated, based on its attack of other nearby locations. Furthermore, because the Israelite chariot corps was incorporated into the Assyrian army, most of the equipment logically went with the horses.[99]

Although the absence of horse trappings is somewhat puzzling, it is not conclusive. Given the high value of trained chariot horses; it makes sense that, before the attack on Northern Israel and Megiddo by Tiglath-pileser III in 732 B.C.E., the horses in training at Megiddo were evacuated to safer places to avoid capture by the Assyrians. In order to move the horses, the keepers would have had to move their bridles, bits, harnesses, and other "trappings" with them; these items were probably custom-made and handcrafted to fit each horse.[100] If the Megiddo stables were, in fact, evacuated prior to the Assyrian invasion, one would not expect to find any useful objects relating to the horses in the area. Obviously, if items such as bridles and bits were left behind inadvertently, they would have been confiscated by the Assyrians or resmelted into tools by later occupants. Megiddo was not a city that sat idle for 2,000 years, waiting to be excavated; it was a city that was rebuilt time and again by occupying forces.

Regarding the absence of bits, we must not presuppose that the training protocol at Megiddo even required the use of bits. Horse professionals tend to agree that control of a horse has more to do with the training and expertise of the rider than with the bitting. Aelian described the superiority of Indian horsemanship and noted that the horses were not bitted; rather, they wore spiked muzzles: "thus their tongue goes unpunished and the roof of their mouth untormented."[101] He further noted that accomplished equestrians had enough control of the horses to drive chariots in

97. D. Dunham, *The Royal Cemeteries of Kush*, vol. 1: *El Kurru* (5 vols.; Cambridge: Harvard University Press, 1950) 110; V. Karageorghis, "Horse Burials on the Island of Cyprus," *Arch* 18 (1965) 282–90; Schauensee, "Horse Gear," 39–51.

98. Although no chariot burials have been discovered in ancient Israel, there is an oblique reference in Isa 22:18 to death in a chariot associated with the tomb of Shebna, a proud servant of King Hezekiah.

99. S. Dalley and J. N. Postgate, *The Tablets from Fort Shalmaneser* (Cuneiform Texts from Nimrud 3; London: British School of Archaeology in Iraq, 1984) 35 and tablet 99:167–79.

100. The importance of correctly fitting bridles and bits remains a significant concern of equestrians today. Because horse heads and muzzles vary substantially in size, the factory-made, mass-produced bridles in modern times are designed with buckles so that they may be individually fitted and adjusted to each horse. Bridles in antiquity may also have been adjustable, because a well-fitting bridle was essential for controlling a horse.

101. Aelian, *On the Characteristics of Animals* (trans. A. F. Scholfield.; 3 vols.; LCL; Cambridge: Harvard University Press, 1959–71) 13.91.

precise circular formations. Likewise, Native American horsemen, renowned for their equestrian prowess, did not use metal bits. Their horses were bitted with a "thong," which was nothing more than a braided piece of leather or rawhide that went over the tongue and was tied under the horse's chin.[102] Western-style riders today often use hackamores (halter-like bridles without bits), rather than bitted bridles, on their horses. The horses are customarily taught to "neck rein," to move away from the touch of the rein on their neck, rather than the pressure of a bit.

Further, it is not correct to assume that the Megiddo stables are totally devoid of horse-related objects dating to the Iron Age strata. A clay horse head with an incised harness that served as the pouring spout of a zoomorphic vessel was excavated from the Northern Stables (Area L) in 2002. Two somewhat similar clay horse heads with harnesses were discovered in the Northern Stables in 2004. Additionally, a small clay horse head wearing a decorated brow band and blinkers that may have been a part of a horse figurine was found in Area K.[103] The original purpose of these finds is unknown, although we may speculate that the zoomorphic vessels with an open pouring spout found in the stables could have contained olive oil for application to the horses' hooves (see fig. 5.7).

Toggles—small, smooth-rounded oval stones with incised grooves, similar to those found at Fort Shalmaneser in Nimrud—are also included among the small finds at Megiddo, although they were not identified as being horse related.[104] Toggles were useful in properly adjusting the tension of the straps that secured the leather saddle blankets on cavalry and chariot horses. A rough comparison may be made to the leather and metal cinching device on modern-day Western saddles. More research is warranted on the prevalence and context of toggles at Megiddo and other sites in Israel.

In addition, an interesting jar handle from Iron Age Megiddo bears a rectangular impression of a horse with a suckling foal.[105] The foal may be identified by its placement under the mare and by its short, upraised tail, which is often seen when foals are feeding.

### *Lack of Horse Bones at Megiddo*

To horse professionals, it is difficult to understand why anyone would expect to find horse bones in a stable. Life-threatening accidents and injuries such as broken

102. G. Pony Boy, *Horse, Follow Closely: Native American Horsemanship* (Irvine, CA: BowTie, 1998) 54.

103. B. Sass and G. Cinamon, "The Small Finds," in *Megiddo IV: The 1998–2000 Seasons* (ed. I. Finkelstein, D. Ussishkin, and B. Halpern; 2 vols.; Monograph Series 24; Tel Aviv: Tel Aviv Institute of Archaeology, 2006) vol. 2, fig 18.39(804), p. 409; fig. 18.46(1016–1017), p. 424; fig.18.38(797), p. 408.

104. Curtis and Reade, *Art and Empire*, 168. See Lamon and Shipton, *Megiddo I*, pl. 107, no. 13–15; I. Finkelstein, D. Ussishkin, and B. Halpern, *Megiddo III* (2 vols.; Monograph Series 18; Tel Aviv: Institute of Archaeology, 2006) 1:417, fig. 12.53, no. 10.

105. Ibid., 1:411, fig. 12.47, no. 3.

legs and lightening strikes generally occur in pastures, not stables. Horses can live into their late 20s or early 30s.[106] Generally, older horses are turned out to pasture, where they die from natural causes, and scavengers devour their bodies and disperse their bones.[107] Few horses die in their stalls unless there is a fire in the barn. Sick horses display warning signs and are typically removed from the stable area immediately if they appear to be suffering from a contagious disease. At the first signs of colic, the most frequent cause of death among stabled horses, a horse is taken out of the stall and hand-walked until the vet arrives. If it appears that the horse is going to die, every possible effort is made to move it while still alive to a pasture where its carcass can be burned, buried, or hauled away. The *last* place a horse professional wants a horse to die is in the stable, where its removal is difficult, and its death is disturbing to the other horses. The only place that horse bones should logically be expected is in ceremonial horse burial sites, but in a stable, only in the event of a flash fire such as the fire at Hasanlu.

As I suggested above regarding the lack of archaeological evidence for the Roman cavalry, it is likely that a special horse-disposal/recycling center was established outside the compound for the processing of the carcasses into meat, leather, and other by-products.[108] In particular, it is feasible that the bones were boiled for the oil they contained, which was also the practice in World War I. During the British cavalry expedition into Iraq in 1914–18, dead horses were disposed of by tossing them into the river, because the shortage of wood prevented burning them.[109] Regardless, it is important to reiterate that the death of a horse at a training center, as compared with a battlefield, would be an infrequent occurrence.

## *Lack of Horse Teeth at Megiddo*

There are several reasons that the lack of horse teeth at or near the Megiddo stables is not a determining argument in the stable debate. First, the stables were originally excavated before the introduction of modern archaeozoology, and earlier-discovered teeth may not have been considered significant enough to note. Second, the stables were emptied before the Assyrians tore down their brick walls. Third, assuming that the Megiddo site was used primarily as a training facility, it would not necessarily be expected to yield teeth during excavation. A horse begins to lose its milk teeth (12 incisors and 12 molars) when it is two and one-half years old, shedding the center

106. Aelian recorded that the life-span of stallions was 35 years but noted that offspring born to old horses tended to be feeble and weak in the legs (*On the Characteristics of Animals* 15.251).

107. Upon his return from the Megiddo 1998 Expedition where the "absence-of-bones" issue was debated, Tom Moon reported that once, he disposed of a horse carcass in a ravine on his farm in Williamson County, TN. Within three months scavengers had disposed of the carcass and bones completely (private communication).

108. Dixon and Southern, *The Roman Cavalry*, 178–80.

109. Anglesey, *A History of the British Cavalry*, 50. A biblical reference to boiling, seething, and charring animal bones is found in Ezek 24:4–5, 10.

incisors first. At age four, the intermediate teeth are shed, and at age five, the outer incisors are lost. The horse has its "full mouth" of permanent teeth by age five.[110]

Coincidently, it is at about age five that the performance horse's serious training (requiring a stabling regimen, as opposed to general education) usually begins.[111] This is not to say that only five-year-old horses were present at Megiddo; however, if it functioned as a chariot-training center, it is likely that most of the horses in training were at least five years old and past the age of shedding teeth. Obviously, if younger horses shed their teeth in the feeding troughs, the teeth would have been removed by the grooms to prevent the horses from swallowing them with their grain; but it is more than likely that the teeth of the younger horses were shed in the pastures while grazing. Whatever the scenario, the likelihood of finding horse teeth in the Megiddo stables is very remote.

### *Ventilation*

An essential characteristic of any stable is proper ventilation. Stables can easily become incubators for disease-spreading bacteria that can decimate the horse population. One solution available at the Megiddo stables was to leave part or all of the center aisle open to the sky. The roof over the stall area could have been tall enough (and strong enough) to use as a hayloft, allowing the hay to be dropped into the horses' troughs below. Generally, the ceiling above the horses, if made of wood, needs to be tall enough (approximately 8 ft. or more) to prevent a horse from chewing the wood. Tall ceilings with an open center aisle would catch the breezes blowing across the mound, providing shade and beneficial fresh air for the horses.[112]

### *Training Considerations*

In addition to the vast, flat plain around Megiddo, which was especially suitable for chariotry training, what must have been a state-of-the-art training facility was designed by the architects inside the city. The Southern Courtyard, measuring 60 m square, and plastered with a smooth lime overcoat provided an optimal flat surface suitable for chariotry training for young horses and for teaching advanced maneuvers to the older ones. The plastered surface prevented the area from becoming rutted by the chariot wheels and ground into dust by sharp hooves (see fig. 5.8).

The builders of the Southern Courtyard took great care to ensure a smooth, almost level surface, using arduous ground-fill techniques (up to 4 m thick), extensive mud-brick retaining walls, and drainage systems to facilitate water runoff. The gentle

110. R. E. Taylor and T. G. Field. *Scientific Farm Animal Production: An Introduction to Animal Science* (7th ed.; Upper Saddle River, NJ: Prentice Hall, 2001) 598.

111. Roman racehorses did not compete before age five; Hyland, *Equus*, 46. Furthermore, in modern times, the rules for Olympic equestrian sports require the horses to be at least seven.

112. Belkin and Wheeler, "Reconstruction," 679–86.

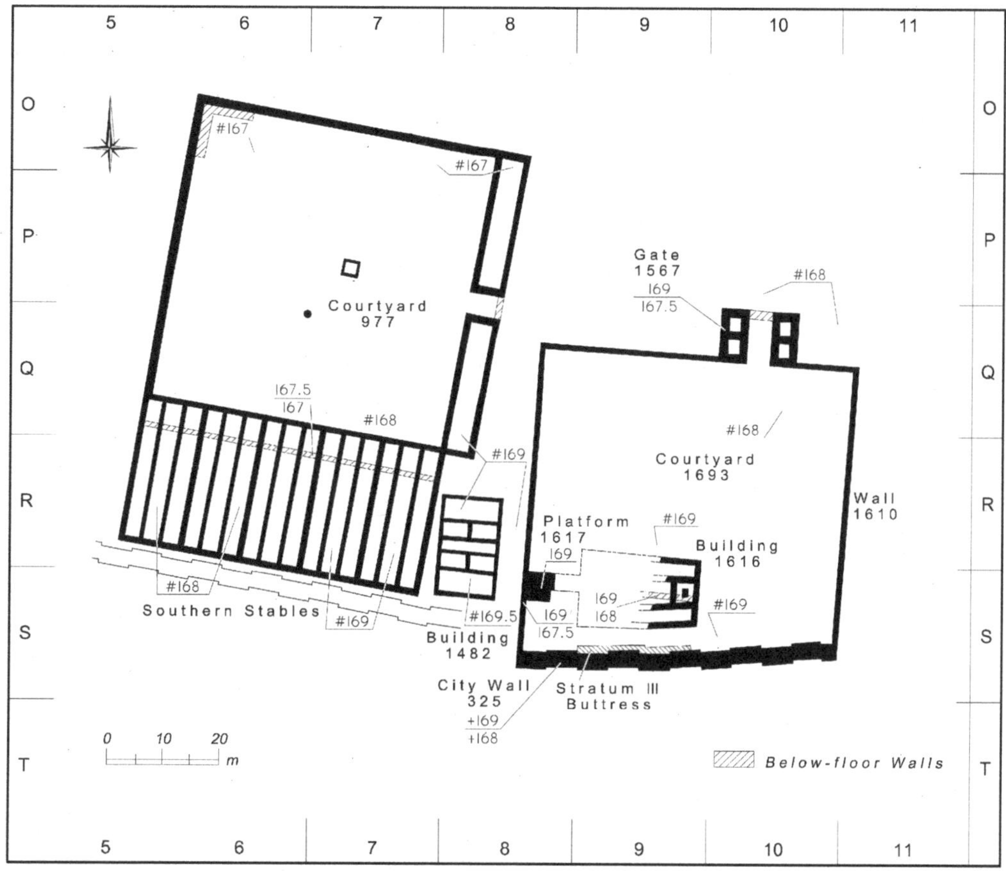


*Plan of Area A, Stratum IV.*

Fig. 5.8. Megiddo, Southern Stables and courtyards. Plan from Norma Franklin, *State Formation in the Northern Kingdom of Israel: Some Tangible Symbols of Statehood* (Ph.D. dissertation, Tel Aviv University, 2006). Published with the permission of Norma Franklin.

slope of about 1.5 m toward the north gave the horses the simulated training benefit of pulling against the yoke on undulating terrain. The five stable units that opened onto the courtyard had entrances flush with the courtyard floor that enhanced the ease of ingress and egress for the horses.

Teaching a horse balance and obedience is the most important aspect of training for pulling wheeled vehicles, whether ancient chariotry or modern driving. If a horse cannot balance itself in the turns, it will go down and often flip the conveyance that it is pulling.[113] Good balance must be achieved before speed is encouraged. The

113. S. Walrond, *A Guide to Driving Horses* (Hollywood: Wilshire, 1971) 71; D. L. Ganton, *Drive On: Training and Showing the Advanced Driving Horse* (Hollywood: Wilshire, 1982) 106.

enclosed wall around the Southern Courtyard easily served the purpose of containing the horses as well as providing four safe corners in which to practice balancing in the turns. Furthermore, the enclosed arena prevented the horses from escaping and running away if they panicked during the training process.

### *Jezreel Military Headquarters and Cavalry Depot*

Assuming that the Megiddo training complex and stables date to the eighth century, the question arises where the Omride facility for training the chariot army of Ahab was located. An inviting answer is to place it at the Jezreel fortress located only eight miles from Megiddo. The Jezreel compound was an eleven-acre enclosure situated on the prominent summit of a ridge overlooking the Valley of Jezreel and the highway from Megiddo to Beth-shean. The Jezreel fortress had casemate walls, a rock-cut moat, and a six-chambered gate at its entrance. The inside of the enclosure had few buildings, but the excavation showed that a substantial area had been carefully leveled, perhaps for chariot use. Tripartite pillared buildings have not been discovered at Jezreel, although only a very small part of the site has been excavated. The towers at the corners of the Jezreel fortress offered 360-degree visibility, making it highly unlikely that an ambush of the facility would have been possible. Jezreel's strategic location, easy access to ample supplies of pasture and grain, the large spring at the base of the fortress, and the large open area inside the enclosure certainly made it the perfect place to prepare horses for war.

According to Israeli archaeologist David Ussishkin, who conducted excavations there between 1990 and 1996 C.E., the fortress was built in the early ninth century and was destroyed by Hazael and the Aramean army in the later part of the ninth century.[114] Ussishkin suggests that the Jezreel enclosure served as the central base for the cavalry and chariotry units of the Israelite army under Ahab.[115] Of course, this does not mean that all of the horses were stabled there; it may have functioned primarily as a mustering center.

Given the suitability of northern Israel for horses and chariots, not only the Jezreel Valley but Iron Age facilities such as Megiddo, Hazor, Dan, Kinrot, Beth-shean, and Bethsaida could be considered centers that played a part in an extensive horse-management program during the ninth and eighth centuries. It is possible that the entire Northern Kingdom was in effect one large horse depot and that Jezreel was

114. D. Ussishkin, "Samaria, Jezreel and Megiddo: Royal Centres of Omri and Ahab," in *Ahab Agonistes: The Rise and Fall of the Omri Dynasty* (ed. L. L. Grabbe; European Seminary in Historical Methodology 6; London: T. & T. Clark, 2007) (1997) 293–309. See also N. Na'aman, "Historical and Literary Notes on the Excavation of Tel Jezreel," *TA* 24 (1997) 122–28.

115. D. Ussishkin, "Excavations at Tel Jezreel 1992–1993: Second Preliminary Report," *Levant* 26 (1994) 1–48; contra this dating, see N. Franklin, "Jezreel: Before and after Jezebel," in *Israel in Transition: From Late Bronze II to Iron IIA (c. 1250–850)* (ed. L. L. Grabbe; London: Continuum, 2008).

the equivalent of an Omride "Fort Shalmaneser" with its parade ground, assembly area, and chambered gates—a prototype for what was later built at Megiddo. If so, the horses did not need to be headquartered at Jezreel year-round; they might have been brought in for training and redispersed until needed. It is reasonable to believe that soldiers and horses housed within the Jezreel compound were safely removed immediately prior to invasions and attacks. The horses would probably have been evacuated to Megiddo, Samaria, and even Judah.

Jezreel is mentioned with more specificity in connection with cavalry horses, chariots, and military matters than any other location in the Bible. For example, the Jezreel fortress apparently served as a facility where injured kings could safely retire to heal or regroup from battle wounds (2 Kgs 8:29, 9:16). The plain near Jezreel was the scene of Jehu's coup with its frenzied chariot driving and the deaths of King Joram and Ahaziah by chariot archers (2 Kgs 9:14–28). Jezreel was also the scene of Queen Jezebel's death by the trampling of Jehu's horses (2 Kgs 9:33).

## *Summary*

To the horse professionals who worked at or visited Megiddo in the 1998 and 2000 seasons, it was quite clear that ancient architects gave considerable thought to the design of its stables. Many features, such as the center aisle, troughs, and pillar dividers are still in use in modern barns 2,800 years later. The fact that horse safety was also a design concern is apparent from the raised stone troughs, the tethering holes, and the single-exit stable. The support systems for the stable facilities, which include the grain silo and water system, appear more than adequate to allow the stabling of approximately 450 horses. It is certain that the stables were designed to protect valuable animals while providing abundantly for their maintenance.

It is safe to say that Megiddo was a city dedicated in various degrees to the care and training of horses from the time of Thutmose III (1479) until well after its capture by the Assyrians in 732. It appears to have been the nerve center of Israel's defense system, which during most of the Iron Age depended on chariotry training. It is also quite plausible that the horses trained at Megiddo were sold or traded to neighboring countries during the eighth century.[116]

116. For a discussion of the day-to-day operational requirements of the Megiddo chariotry facility and the historical context, see Cantrell and Finkelstein, "A Kingdom for a Horse," 643–65.

## Chapter 6

# *Warfare in Iron Age Israel*

*Their arrows are sharpened,*
*and their bows are drawn.*
*Their horses' hoofs are like flint,*
*their chariot wheels like the whirlwind.*
*(Isa 5:28)*

Chariot warfare crystallized in ancient Israel long before the Iron Age. The earliest chariot battle recorded in detail was fought at Megiddo in 1479 between Pharaoh Thutmose III and a Canaanite coalition. The battle resulted in the Egyptian king's capturing 924 chariots and over 2,238 horses.[1] His son, Amenhotep II (ca. 1447–1421), also battled the Canaanites in the Jezreel Valley (ca. 1430) and captured 820 horses and 730 chariots.[2] During the Amarna period, Biridiya, the ruler of Megiddo, wrote letters (ca.1360–1350) to Amenhotep III and Akhenaten requesting Egyptian chariots to fight his enemy, who was capturing towns in the Jezreel Valley and leaving him stranded in Megiddo.[3] Almost 100 years later, Ramesses II rolled his chariotry through the Levant to engage the Hittites at the Battle of Qadesh on the Orontes (1285). This great chariot battle involved thousands of chariots and was claimed as a victory on both sides.[4]

1. J. K. Hoffmeier, trans., "The Annals of Thutmose III," *COS* 2:2:7–13. The total number of 2,238 is divided into 2,041 horses, 191 foals, and 6 stallions, with the number of colts not provided. For the suggestion that the mares and stallions captured at Megiddo launched a mass breeding policy in Egypt, see A. Hyland, *The Horse in the Ancient World* (Westport, CT: Praeger, 2003) 80–83.

2. J. K. Hoffmeier, trans., "The Memphis and Karnak Stelae of Amenhotep II," *COS* 2:3:19–23. For a discussion of the Egyptian battle tactics around Megiddo and the Jezreel Valley, see E. H. Cline, *The Battle of Armageddon: Megiddo and the Jezreel Valley from the Bronze Age to the Nuclear Age* (Ann Arbor: University of Michigan Press, 2000) 32–37.

3. For a description of the usage of chariotry at Megiddo and nearby during the Amarna period, see B. Halpern, "Centre and Sentry: Megiddo's Role in Transit, Administration and Trade," in *Megiddo III: The 1992–1996 Seasons* (ed. I. Finkelstein, D. Ussishkin, and B. Halpern: 2 vols.; Monograph Series 18; Jerusalem: Tel Aviv University Press, 2000) 2:535–77. See also Cline, *Battle of Armageddon,* 37–43.

4. M. Lichtheim, *AEL*, 2:57–72; A. J. Spalinger, *War in Ancient Egypt: The New Kingdom* (Malden, MA: Blackwell, 2005) 149. For a detailed assessment of the comparative performance of the lightweight

Early in the Iron Age, the Egyptian chariotry, led by Shoshenq (Shishak I, 945–924) again invaded Israel (ca. 925); this invasion led to the capture and subjugation of Megiddo, Gibeon, and numerous other cities.[5] The biblical text records that Shoshenq arrived with 1,200 chariots and 60,000 "horsemen" (2 Chr 12:3).[6] Shoshenq erected a victory stele at Megiddo, just as Thutmose III had done at the site 500 years earlier.[7] It was after Shoshenq returned to Egypt that Israel and Judah fortified their cities to provide defenses against chariotry (1 Kgs 15:16–22, 2 Chr 14:6–7). Both nations eventually developed their own standing armies and large, highly organized chariot forces.

Israel's chariotry did not go untested. In fact, Israel's commitment to building an extensive infrastructure to support its chariotry was essential to its continued existence as a nation. Military historian Steven Weingartner explains, "More than just an effective tactical, or battlefield, weapon, chariotry was the key to building and maintaining empires." Quite simply, both Israel (and Judah) understood that "it took a chariotry to fight a chariotry."[8] Israel's fractious neighbors Egypt, Assyria, and the Arameans had their own highly developed chariotries and cavalries, which they deployed frequently to challenge Israel's sovereignty.[9]

---

Egyptian chariots against the heavier Hittite chariots, see R. C. Suhr, "Ambush at Kadesh," *Great Battles* (2006) 16–21.

5. I. Finkelstein, "The Campaign of Shoshenq I to Palestine: A Guide to the 10th Century B.C.E. Polity," *ZDPV* 118 (2002) 109–35; N. Na'aman, "The Northern Kingdom in the Late Tenth–Ninth Centuries B.C.E.," in *Understanding the History of Ancient Israel* (ed. H. G. M. Williamson; Proceedings of the British Academy 143; Oxford: Oxford University Press, 2007) 399–418. See also J. S. Holladay Jr., "The Kingdoms of Israel and Judah: Political and Economic Centralization in the Iron IIA–B (ca. 1000–750 B.C.E.)," in *The Archaeology of Society in the Holy Land* (ed. T. E. Levy; London: Leicester University Press, 1998) 372–75.

6. Egyptians are usually depicted driving chariots and almost never riding horses, thereby making the translation of *pārāšîm* פרשים as 'horsemen' doubtful; a better translation is 'horses', as it is translated in the LXX. Furthermore, "60,000" makes no sense, although 6,000 horses accompanying 1,200 chariots would be appropriate. Notably, the earlier, more abbreviated account of Shoshenq's invasion found in 1 Kgs 14:25–28 does not specifically mention a chariot invasion.

7. R. B. Partridge, *Fighting Pharaohs: Weapons and Warfare in Ancient Egypt* (Manchester, UK: Peartree, 2002) 280–81. See also *ANET*, 263–64. Some scholars suggest that Shoshenq did not destroy Megiddo prior to erecting the victory stele because he intended to use Megiddo as an Egyptian administrative center. They argue that the site was destroyed earlier by other forces, either seismic or military; see I. Finkelstein, D. Ussishkin, and B. Halpern, "Archaeological and Historical Conclusions," *Megiddo IV: The 1998–2002 Seasons* (ed. I. Finkelstein, D. Ussishkin, and B. Halpern; 2 vols.; Monograph Series 24; Tel Aviv: Tel Aviv Institute of Archaeology, 2006) 2:843–60. See also I. Finkelstein, "The Campaign of Shoshenq I to Palestine."

8. S. Weingartner, "Chariots Changed Forever the Way Warfare Was Fought, Strategy Conceived, and Empires Built," *Military Heritage* (August 1999) 19–27. For a discussion of chariots as strategic weapons in sieges and as mobile defense units, see R. A. Gabriel, *Military History of Ancient Israel* (Westport, CT: Praeger, 2003) 23.

9. Philip J. King and Lawrence E. Stager, *Life in Biblical Israel* (ed. D. A. Knight; Library of Ancient Israel; Louisville, KY: Westminster John Knox, 2001) 244–47; Y. Yadin, *The Art of Warfare in Biblical Lands* (2 vols.; New York: McGraw-Hill, 1963) 1:284–87, 2:297–313.

Israel could not have prevented the repeated attempts at invasion and conquest without a formidable chariotry to patrol its borders. During the Monarchic period, the country had the physical and human resources as well as the leadership and organizational skills to field a significant army and a sophisticated chariotry. Israel's strategic geographical location made it the virtual hub of chariot warfare in the Iron Age. In fact, during this period, Israel was invaded by the chariot forces of Aram, Assyria, Moab, and Egypt.[10] According to the biblical text, chariot battles were fought at or near Hazor, Megiddo, Gilgal, Mt. Gilboa, Helam, Samaria, Dothan, Aphek, Ramoth-gilead, Jezreel, and in Moab and Edom.[11] Ironically, Saul (ca. 1021–1000), the first king of the United Monarchy, was said to take his own life because he had no chariot or horse to escape the Philistine chariot archers who pursued him up Mount Gilboa (2 Sam 1:6–10).[12] Chariot archers were also responsible for the battle deaths of other Israelite and Judahite kings: Ahab, Joram, Ahaziah, and Josiah (1 Kgs 22:31–37; 2 Kgs 9:21–24, 27; 23:29).

Israel's geographical position, precariously lodged between countries with long histories of sophisticated, highly organized armies and experienced chariotry units made it paramount that the Israelite chariotry be swiftly deployable at the first sight of invaders. The six-chambered gates at fortress entrances made rapid deployment possible. It may be argued that the Israelite chariotry enjoyed such a profound military advantage operating on its home turf with conveniently located defensive fortresses, skilled horsemen, and highly trained horses that its enemies were hesitant to invade until they could arrive with chariotry and cavalry forces that far outnumbered those of Israel. From the Egyptian invasion by Shishak (925) until the Assyrian campaign by Tiglath-pileser III (734), Israel organized and developed exceptional chariotry and cavalry divisions that engaged in nearly 200 years of effective border defense. Otherwise, the Assyrians or the Egyptians at their leisure could have conquered the land with little effort. By the strength and skill of the horsemen of Israel, the nation survived. Only the Arameans, who are historically regarded as extremely skilled horsemen and charioteers, were successful in achieving temporary control of key locations in Israel in the ninth century (840–800), when they occupied Hazor and Dan and

10. For details of the Moabite invasion of Israelite locations in the Transjordan, see A. F. Rainey and R. S. Notley, *The Sacred Bridge: Carta's Atlas of the Biblical World* (Jerusalem: Carta, 2006) 203–5.

11. Hazor (Waters of Merom) Josh 11:1–19; Megiddo (Kishon River), Judg 4:1–16; Gilgal (Michmas), 1 Sam 13:5–7; Mt. Gilboa, 2 Sam 1:6; Helam, 2 Sam 11:17–19; Samaria, 1 Kgs 19:20, 20:1; 2 Kgs 6:24, 10:2–17; Dothan, 2 Kgs 6:8–22; Aphek, 1 Kgs 10:26–34; Ramoth-gilead, 1 Kgs 22:29–37; Jezreel, 2 Kgs 9:16–34; Moab, 2 Kgs 3:1–26; and Edom, 2 Kgs 8:21–22. For a discussion of battle strategies and the potential effectiveness of the Israelite chariotry, see C. Herzog and M. Gichon, *Battles of the Bible: A Military History of Ancient Israel* (New York: Random, 1978) 93ff. See also Yadin, *Art of Warfare*, 2:297–313.

12. For a discussion of the various military options in this battle, see Gabriel, *The Military History of Ancient Israel*, 216–20.

destroyed towns in the Galilee and Jezreel Valley, including Megiddo, Jezreel, and Taanach.[13]

## *Aramean Invasions*

Aramean, Assyrian, and Hebrew texts and steles indicate that the Arameans had thousands of horses and routinely engaged in chariot warfare and mounted combat. According to 2 Sam 8:4 (NIV, LXX), David (ca. 1000 B.C.E.) captured 1,000 chariots and 7,000 charioteers (perhaps horses) from Hadadezer, king of Aram.[14] The Arameans also hired out their chariots and charioteers as mercenaries. There is a tradition in the Hebrew Bible that some 700 Aramean charioteers were killed while fighting as mercenaries for the Ammonites in a battle against David and the United Monarchy (2 Sam 10:18). Furthermore, an Aramean unit of chariotry in the Assyrian army is described in the horse lists found at Fort Shalmaneser, although there is no information about the date or origin of its formation. Dalley and Postgate tentatively suggest that the Aramean chariotry unit was formed during the reign of Tiglath-pileser III (745–727).[15]

At the Battle of Qarqar (853), Adad-idri of Aram–Damascus (880–842) contributed 1,200 chariots and 1,200 cavalry (3,600 to 4,800 horses)—the next-largest presence after Ahab—to the coalition forces fighting the Assyrian army.[16] In 849, 848, and 845, Hadadezer, the Aramean king, led a coalition force, including Irhuleni of Hamath and the "twelve kings of the sea coast," against the invasion forces of Shalmaneser III.[17] Shalmaneser III claims to have defeated the coalition armies and captured their chariots, cavalry horses, and battle equipment.[18] According to the Assyrian annals, Shalmaneser III also captured 1,121 chariots (2,000 or 3,000 horses) and 470 cavalry

13. H. M. Niemann, "Core Israel in the Highlands and Its Periphery" in *Megiddo IV: The 1998–2002 Seasons* (ed. I. Finkelstein, D. Ussishkin, and B. Halpern; 2 vols.; Monograph Series 24; Tel Aviv: Tel Aviv Institute of Archaeology, 2006) 2:821–42.

14. N. Na'aman suggests that the figure of Hadadezer was modeled upon the figure of Hazael and that the narrative more likely reflects ninth-century events ("In Search of Reality behind the Account of David's Wars with Israel's Neighbors," in *Ancient Israel's History and Historiography: The First Temple Period* [*Collected Essays* 3; Winona Lake, IN: Eisenbrauns, 2006] 38–61). Contra: Stuart A. Irvine, "The Last Battle of Hadadezer," *JBL* 124 (2005) 341–47.

15. S. Dalley and J. N. Postgate, *The Tablets from Fort Shalmaneser* (Cuneiform Texts from Nimrud 3; London: British School of Archaeology in Iraq, 1984) 32, 36.

16. These numbers appear on the Kurkh Monolith; S. Yamada, *The Construction of the Assyrian Empire: A Historical Study of the Inscriptions of Shalmaneser III (859–824 B.C.) Relating to His Campaigns to the West* (Culture and History of the Ancient Near East 3; Leiden: Brill, 2000) 156.

17. For a discussion of the scholarly identification of Hadadezer with Ben-hadad, see K. L. Younger Jr., "Neo-Assyrian and Israelite History in the Ninth Century: The Role of Shalmaneser III," in *Understanding the History of Ancient Israel* (ed. H. G. M. Williamson; Proceedings of the British Academy 143; Oxford: Oxford University Press, 2007) 243–77.

18. Rainey and Notley, *Sacred Bridge*, 201.

from the Aramean king Hazael (842–796) in 841.[19] Regardless of whether the numbers of horses and chariots were routinely inflated, as believed by many authorities, it is clear that the Arameans had a culture of warfare in which horses and chariots played a paramount role.[20] They dominated the Northern Kingdom of Israel in the second half of the ninth century, conquering Dan and Hazor, and revamped a large fortress with chambered gates at Bethsaida.[21]

Concerning the accounts of the Aramean invasions in the Hebrew Bible, Nadav Na'aman points out that scholars generally believe that the battle stories between Aram and Israel were inserted into the book of Kings by a post-Deuteronomistic editor. As Na'aman explains: "However, the majority of these narratives were written earlier than the composition of the Books of Kings, and even if inserted at a later stage, their contribution to the history of the Northern Kingdom must be carefully examined, each story in its own right."[22] He believes that the author of the books of Kings made an effort to assemble all the available sources (now lost) and used them in writing his work. Therefore, they are worthy of investigation, notwithstanding the numerous difficulties with their chronology.

According to the biblical text, the Arameans invaded Israel around 857, when the Aramean king persuaded 32 other kings along with their chariot contingents to besiege and attack the capital in Samaria (1 Kgs 20:1–22).[23] In the face of this sizable display of military strength, Ahab surprised the Arameans while they were sleeping.

19. Yamada, *Construction of the Assyrian Empire*, 229, 234. Additionally, as mentioned above in the discussion about the Israelite breeding program, the 12 years between the Battle of Qarqar and the battle against Hazael were quite adequate for the Arameans to replace their chariotry, assuming an ongoing breeding and training program.

20. Younger, "Neo-Assyrian and Israelite History," 243–77; and A. Kirk Grayson, "Shalmaneser III and the Levantine States: The 'Damascus Coalition,'" *JHS* 5 (2004) Article 4, pp. 1–10, www.arts.ualberta.ca/JHS/Articles/article_34.htm.

I am reluctant to discount these numbers as being grossly inflated for two reasons: first, I find it too much of a coincidence that three separate sources from three distinct cultures so unlikely to have consulted one another all place the calculations of chariots and horses in the thousands during the Iron Age in and around Israel; and second, from a horse-management perspective, there was plenty of pasture, grain, and free labor for each nation to support between 5,000 and 10,000 horses in order to field 1,500 to 2,000 chariots routinely. Furthermore, as a practical matter, after a battle it was important to note with some exactitude how many horses were captured in order to know how much food to requisition for their immediate needs; it is not implausible that, later, these numbers made it into the official records and annals.

21. I. Finkelstein, "Hazor and the North in the Iron Age: A Low Chronology Perspective," *BASOR* 314 (1999) 55–70; R. Arav and R. A. Freund, eds., *Bethsaida: Bethsaida Excavations Project Reports and Contextual Studies* (4 vols.; Kirksville, MO: Thomas Jefferson University Press, 1995–2009) 3:13.

22. Na'aman, "The Northern Kingdom," 408–9.

23. Rainey and Notley, *Sacred Bridge*, 199. Some scholars date the events of 1 Kings 20 to the time of Jehu (841–814); see Na'aman, "The Northern Kingdom," 408. Others date these events to the time of Joash (798–782); see E. Lipiński, *The Aramaeans: Their Ancient History, Culture, Religion* (OLA 100; Leuven: Peeters, 2000) 397.

The king of Aram and some of his officers escaped on horseback, but Ahab captured the horses and chariots left in the camp (1 Kgs 20:21). If so, many of these captured horses undoubtedly joined Ahab's chariotry ranks and likely participated in the Battle of Qarqar four years later, a battle in which the Arameans were members of the allied forces fighting the Assyrians.[24]

The next spring, the Arameans returned and lured the Israelite army to fight an open-terrain battle at Aphek, approximately 30 miles from the military compound at Jezreel. The biblical account mentions the muster by both countries prior to the battle. The Israelite army is described as being camped like "two small flocks of goats" opposite the Arameans, who covered the countryside (1 Kgs 20:27).[25] It was a victorious chariot battle for the Israelites; however, the only details provided by the author involve the death of Aramean foot soldiers and a peace treaty made in Ahab's chariot (1 Kgs 20:29–34). Despite the lack of clarity in the text, the short distance from the Israelite military headquarters at Jezreel (30 miles) and Beth-shean (25 miles) to the battlefield at Aphek would have made providing fresh horses, soldiers, and supplies relatively manageable. As a practical matter, the Israelites could replenish their army within a few hours and had no need to station their entire force at the battle campsite.[26]

Although the king of Aram had agreed in a covenant to return all the cities that belonged to Ahab, king of Israel (1 Kgs 20:34), Ramoth-gilead remained in Aramean possession. Israel and Judah combined their chariotries to attack the Arameans at Ramoth-gilead in 852. The Battle of Ramoth-gilead was an unmitigated disaster for Israel and Judah, resulting in the death of Ahab and the near death of Jehoshaphat (873–849), king of Judah (1 Kgs 20:29–37).[27]

However, when Jehoshaphat initially joined forces with Ahab to reclaim Ramoth-gilead, Jehoshaphat pledged his loyalty by saying, "my horses are as your horses" (1 Kgs 22:4). Taken literally, Jehoshaphat could have moved his horses from Jerusalem or elsewhere in Judah to a mustering center at either Jezreel or Beth-shean in only a

24. In fact, many of the same horses were probably moving from Israel to Aram to Assyria, almost "round-robin" style, based on who captured them at the last battle. Certainly, this could have been the case in the numerous battles between the Israelites and the Arameans in the ninth century.

25. The biblical text also notes that the armies camped seven days before joining battle, perhaps denoting an accepted convention of resting horses and men before a pitched battle, as discussed above in chap. 4.

26. For discussion of the ninth-century administrative centers at Megiddo and Jezreel and the Galilee (Kinneret, Hazor, and Dan), see H. Niemann, "Royal Samaria: Capital or Residence?" in *Ahab Agonistes: The Rise and Fall of the Omri Dynasty* (ed. L. L. Grabbe; Library of Hebrew Bible/Old Testament Studies 421; European Seminar in Historical Methodology 6; London: T. & T. Clark, 2007) 184–207.

27. Naʾaman argues that Ahab was actually killed in the Battle of Qarqar in 853 and that the shift of the setting of Ahab's death from the Assyrians to the Arameans was a result of confusion during the long oral tradition ("The Northern Kingdom," 410–12). If so, it is puzzling that the Assyrians never mention an accomplishment of this sort.

few hours.[28] From Beth-shean to the battlefield at Ramoth-gilead was only another hour or so. The infrastructure was in place to support the horses and the distance to the battlefield reasonable. Presumably neither king would have entered battle without adequate horses and reserves. Alliances and coalitions between Judah and Israel were not strategically difficult because their horses lived in proximity to each other, and both armies could be deployed quickly.

Nevertheless, the Battle of Ramoth-gilead, as told in the Hebrew Bible text, is especially interesting because it plainly illustrates the Arameans' superior horsemanship skills and also reveals why chariot warfare prevailed as the chosen convention of warfare for over 1,000 years. Chariot warfare was both dramatic and dangerous. Only during chariot warfare was it possible to kill a king on an elevated platform in front of so many witnesses, thereby instantaneously effecting regime change and irrevocably altering the course of history.[29] The fact that the story involves Ahab's attempt to disguise himself in the midst of battle suggests that indeed the primary goal in ancient warfare was capturing or killing the opposing leader. Given that chariots were capable of swift pursuit and that potentially lethal shots from chariot archers could be expected any time the chariots were within 250 yards of each other, it was not uncommon for a ruler to have to flee danger. A successful escape from the battlefield required that the king have access to horses that were fresher and faster than those of his enemy.

According to 2 Kgs 13:7, the Arameans reduced the Israelite chariotry to 10 chariots, 50 horsemen (or horses), and 10,000 infantry during the reign of Jehoahaz (814–798). Whether this truly reflects the situation in the Northern Kingdom at this time or not, at a minimum it suggests a memory of the military might of the Arameans. It should be noted, however, that the subjugation of the army does not necessarily mean that the broodmares, foals, and reserve horses also came under the control of the Arameans. Assuming that the breeding stock was reserved under Israelite control, the chariotry and cavalry could be built up again in a matter of a few years, perhaps beginning during the reign of Joash (798–782) and certainly during the 40-year reign of Jeroboam II (782–753).

28. The stables at Lachish were expanded in the eighth century to accommodate a chariotry unit comprising 100 horses; D. Ussishkin, "The Assyrian Attack on Lachish: The Archaeological Evidence from the Southwest Corner of the Site" *TA* 17 (1990) 53–86.

29. Sargon II was also killed in battle while campaigning at Tabal, near the Taurus Mountains (Turkey) in 705, despite the fact that it was customary for his chariot to be surrounded by a large contingent of mounted horsemen; Dalley and Postgate, *Fort Shalmaneser*, 39. Sargon's royal cavalry bodyguard consisted of 1,000 horsemen, making his death in battle all the more mysterious. "This unit escorted the king under all circumstances, and did not budge from his side, neither in enemy nor in friendly country"; Tamás Dezsö, "A Reconstruction of the Army of Sargon II (721–705 B.C.) Based on the Nimrud Horse Lists," *SAAB* 25 (2006) 93–140, esp. p. 96.

### *Assyrian Invasions*

During the first millennium, the Assyrians campaigned against their enemies on an annual basis from the time of Ashurnasirpal II (883–859), usually traveling hundreds of miles to stage open-terrain battles or to siege cities. The army probably marched for almost three months to reach Israel's border.[30] Israel and/or Judah were invaded at various times during the Iron Age, by Shalmaneser III in 841, Tiglath-pileser III in 735–732, Shalmaneser V in 727–722, Sargon II between 722 and 712, Sennacherib in 701, Esarhaddon between 669 and 663, and Ashurbanipal in 667.[31] During his reign, Tiglath-pileser III (745–727) captured numerous cities, including Megiddo and Hazor, but did not approach the capital, Samaria. In 722, Samaria fell to the Assyrians.[32] The biblical text claims that Shalmaneser V put Samaria under siege because King Hoshea ceased paying tribute and was sending envoys to the king of Egypt (Soʾ), presumably for help in mounting a rebellion (2 Kgs 17:3–6). However, Sargon's annals claim that he was the conqueror of Samaria: "I besieged and conquered Samarina. I took as booty 27,290 people who lived there. I gathered 50 chariots from them."[33] Although scholars debate whether Shalmaneser V or Sargon II was in power during the fall of Samaria, it is clear that Samaria never recovered its independence and remained an Assyrian province until the end of the empire.[34]

After the fall of Samaria in 720, Sargon II continued his invasions southward to conquer cities in Judah, gain control of Gaza, and defeat Ashdod in 712.[35] In 701, Sennacherib, Sargon's son, returned to attack Judah, conquering 46 cities, including the military headquarters and horse compound at Lachish. He placed Jerusalem under siege:

> [As for] Hezekiah, the Judaean [who had not submitted to my yoke,] I surrounded and conquered 46 of his strongly fortified walled cities and countless small towns in their vicinity by stamping down siege ramps, bringing up battering rams, the relentless attacks of footsoldiers, bored holes, breaches, and picks. I brought out of their

30. See calculations for a marching army covering 25 km daily and 30 km daily in I. Ephʿal, "Warfare and Military Control in the Ancient Near Eastern Empires: A Research Outline," in *History, Historiography, and Interpretation: Studies in Biblical and Cuneiform Literatures* (ed. H. Tadmor and M. Weinfeld; Jerusalem: Magnes / Leiden: Brill, 1983) 88–106, esp. p. 99.

31. Rainey and Notley, *Sacred Bridge*, 208, 228–32, 234–36, 239–45, 247–50.

32. For discussion of dating issues, see Nadav Naʾaman, "The Historical Background to the Conquest of Samaria (720 B.C.E.)," *Ancient Israel and Its Neighbors: Interaction and Counteraction (Collected Essays*; 3 vols.; Winona Lake, IN: Eisenbrauns, 2005–6) 1:76–93; B. Becking, *The Fall of Samaria: A Historical and Archaeological Study* (Leiden: Brill, 1992) 45–60.

33. K. L. Younger Jr., trans., "The Great 'Summary' Inscription," *COS* 2:118E:296–97.

34. Idem, "The Fall of Samaria in Light of Recent Research," *CBQ* 61 (1999) 461–81. See also Gösta W. Ahlström, *The History of Ancient Palestine* (Minneapolis: Fortress, 1993) 521.

35. Younger, "The Fall of Samaria," 471–73. Becking, *The Fall of Samaria*, 55. H. Tadmor, "The Campaigns of Sargon II of Assur: A Chronological-Historical Study (Conclusion)," *JCS* (1958) 77–100.

> midst 200,150 people small and big, male and female, horses, wild asses, donkeys, camels, oxen and sheep without number and I classified [them] as spoil.
>
> [As for] him, I enclosed him like a bird in a cage in the midst of Jerusalem, his royal city. I erected fortresses against him and made it unthinkable for him to go out of the gate of his city.[36]

Jerusalem, however, was not forcibly entered by the Assyrians; instead, Judah's rebellion was crushed and a heavy tribute imposed. Once again, the Assyrians captured the livestock, including the horses and mules. Judah was subservient to Assyria during the reigns of Esarhaddon (681–669) and Ashurbanipal (669–625), for the most part, and the Assyrian army apparently marched through the country without opposition on the way to attack Egypt in 674, 669, 667, and 664.[37]

The success of the Assyrian invasions depended significantly on their extensive chariotry and cavalry forces. Likewise, Israel's ability to thwart or forestall the Assyrian advance necessarily relied heavily on its capacity to field an opposing chariotry of significant size and experience. Both armies valued seasoned battles horses and the charioteers and equestrians who trained them.

The relatively small size of Israel allowed for the effortless and expedient mustering and training of horses. For Assyria, however, the orchestration of chariot invasions into the Levant was complex and required an inordinate amount of time and coordination. The vast distances from the horse compounds at Fort Shalmaneser (Kalkhu, Nimrud), Nebi Yūnus (Nineveh), and Dur-Sarrukin (Khorsabad) to Urartu where the reserve horses were kept or to the breeding centers in the western Zagros Mountains, made the annual muster a lengthy and complicated ordeal.[38] From the Assyrian correspondence and the "Horse Texts" found at Fort Shalmaneser, the "Review Palace" at the Assyrian capital of Kalkhu (modern Nimrud), it appears that the annual muster took the entire month of March.[39] The Assyrians had special officers, the *mušarkisu*,

36. W. R. Gallagher, *Sennacherib's Campaign to Judah* (Studies in the History and Culture of the Ancient Near East 18; Leiden: Brill, 1999) 127.

37. Rainey and Notley, *Sacred Bridge*, 247–49.

38. See J. N. Postgate, "The Economic Structure of the Assyrian Empire," *Mesopotamia* 7 (1979) 193–221 and nn. 19, 38; H. W. F. Saggs, "Assyrian Warfare in the Sargonid Period," *Iraq* 25 (1963) 153. For a diagram illustrating the complexity of the movements of horses within the far-reaching Assyrian Empire, see J. N. Postgate, *Taxation and Conscription in the Assyrian Empire* (Studia Pohl: Series Maior 3; Rome: Pontifical Biblical Institute, 1974) 211. For a review of the details of the Assyrian mustering process, see F. M. Fales "Preparing for War in Assyria," in *Économie antique: La guerre dans les économies antiques* (Entretiens d'archéologie et d'histoire 5; Saint-Bertrand-de-Comminges: Musée archéologique départemental, 2000) 38–51. See also P. Albenda, "Horses of Different Breeds: Observations in Assyrian Art," in *Nomades et sédentaires dans le Proche-Orient ancien: Compte rendu de la XLVI[e] Rencontre assyriologique internationale, Paris, 10–13 juillet 2000* (ed. C. Nicolle; Amurru 3; Paris: Éditions Recherche sur les civilisations, 2004) 321–34.

39. For a detailed description of the mustering compound, Tell Nebi Yūnus, at Nineveh, see G. Turner, "Tell Nebi Yūnus: The *Ēkal Māšarti* of Nineveh," *Iraq* 32 (1970) 68–88.

in charge of the horse levy. The *mušarkisu* and his assistant, the *šaknu ša pētḫalli*, the cavalry commander, were required to travel from village to village, collecting horses from the provinces and transporting them back to the capital.[40]

At least 13 Samarian deportees served as chief officers in Sargon's chariotry forces according to the Nimrud horse lists (ca. 720).[41] One officer, *Nabû-bēlu-ka''in* became a commander and was the provincial governor of Kār-Šarrukīn and later, Arrapha.[42] Undoubtedly these Israelite charioteers were accomplished horsemen, as their inclusion and acquisition of rank in Sargon's home army indicate.

As described by Esarhaddon, the function of the large "Review Palaces" (*ēkal māšarti*) was to assemble and train the horses prior to the annual battle campaigns: "For setting in order the camp, mustering the steeds, the mules, the chariots, the harness, the battle equipment and the spoil of the enemy . . . for exercising the horse [and] for maneuvering the chariots."[43] When all of the Assyrian war-horses, estimated to have been in excess of 3,000 horses, were finally collected at Fort Shalmaneser, they were then assessed for battle readiness.[44] Only half of the horses were deployed on any given campaign, and the rest were designated "of/for the land."[45] Although it is uncertain what this designation means, at a minimum it seems to indicate that the Assyrians were reluctant to risk their entire chariotry in battle at once. This same prudence in preserving horses may also have applied to the Assyrians' actual battle strategy.

In 841, Shalmaneser III mustered his army and horses and invaded northern Syria. His chariots were met by the Aramean army under the leadership of Hazael, king of Damascus. Assyria claimed victory:

> Sixteen thousand and twenty of his fighting men I felled by the sword. I confiscated one thousand, one hundred and twenty-one of his chariots along with four hundred of his cavalry and his military camp.[46]

It is interesting to note that the number of chariots fielded by Hazael (1121) roughly matched the number of chariots of his predecessor, Hadad-idri of Damascus (1200), at the Battle of Qarqar only 12 years earlier. Taking these numbers at face value, I suggest that either Shalmaneser did not confiscate the chariots of Damascus in the Battle of Qarqar as he claimed or that the chariots were replaced fairly easily.

40. Postgate, *Taxation*, 144.

41. See K. L. Younger Jr., trans., "An Assyrian Horse List," *COS* 3:128:279–80 (this text is also known as *TFS* 99).

42. Dezsö, "Reconstruction of the Army of Sargon," 102.

43. Turner, "Tell Nebi Yūnus," 70. See also S. Dalley, "Ancient Mesopotamian Military Organization," in *CANE*, 1:413–22, esp. p. 419.

44. Assyrian records indicate a total of 3,477 horses and mules gathered by Sargon II for one of the years of his Babylonian campaign, ca. 711–708. Dalley and Postgate, *Tablets from Fort Shalmaneser*, 195–200.

45. Fales, "Preparing for War," 45.

46. Rainey and Notley, *Sacred Bridge*, 208.

It is also worthwhile to consider whether the itemizing of "chariots" in Assyrian booty lists necessarily includes the chariot horses. In the text quoted above, the 1,100+ chariots captured required a minimum of between 2,000 and 3,000 horses to operate (if all fielded at once), yet less than 500 "riding" horses were captured. It appears that a plan was implemented to save as many horses as possible from capture. Logic dictates that, when a chariot was in serious danger of falling to the enemy in battle, the horses would (whenever possible) be cut loose and encouraged to run away. Eventually, they would probably return to their nearby camp or a familiar location behind the battle lines where they could be secured by grooms. The capture of the "470 riding horses" noted in Shalmaneser's booty list could indicate the total number of loose horses captured after the battle or the cavalry horses hurriedly left in an abandoned camp.

As a practical matter, excited horses are not easily caught, even by riders on horseback. It is quite plausible that, after any battle, hundreds of horses roamed free—at least temporarily—and were later reclaimed by their proper owners, if not captured by the enemy. As for the chariots, once the horses were cut loose from them, they would be cumbersome to move and, therefore, were customarily left near the battlefield to be captured and recycled by the enemy.

It appears that under Jeroboam II (782–753), Israel regained various territories to the north, including Dan and Hazor, and even exercised control over Damascus (2 Kgs 13:25). Presumably, this shift in control was in large part due to the rebuilding of Israel's chariotry and the development of a state-of-the-art training facility at Megiddo, as well as Assyria's pressure on Damascus. Undoubtedly, there were other less-grand horse compounds and training centers throughout Israel. Dan and Hazor, with their strategic locations, chambered gates, courtyards, and enclosed compounds are obvious candidates, as are Kinrot and Bethsaida. These outposts/fortresses with their food, water, and shelter for the horses were vital to the reestablishment and management of Israel's chariot force.

It seems certain that the resurgence of Israel's horse operations under Jeroboam II did not go unnoticed by the Assyrians, who may have purchased horses in Megiddo during the eighth century.[47] That Assyria continued to have a great need for trained war-horses is evidenced by the required tribute paid in horses from Egypt, Gaza, Judah, Moab, Ammon, and Edom, as recorded in Nimrud Letter 16 (ND 3765; IM 64159) around 720–715.[48] A strong Israelite chariotry could not be tolerated by Assyria, because it posed a serious threat to Assyria's expansionist goals aimed at Phoenicia, Philistia, and Egypt, as well as Israel itself. Furthermore, the Aramean king Rezin

47. D. O. Cantrell and I. Finkelstein, "A Kingdom for a Horse: The Megiddo Stables and Eighth Century Israel," in *Megiddo IV: The 1998–2002 Seasons* (ed. I. Finkelstein, D. Ussishkin, and B. Halpern; 2 vols.; Monograph Series 24; Tel Aviv: Tel Aviv Institute of Archaeology, 2006) 2:643–65.

48. See K. L. Younger Jr., trans., "A Letter Reporting Matters in Kalaḫ (Kalhu)," *COS* 3:96:245, especially n. 1. See also Ahlström, *History of Palestine*, 642, 651 n. 1.

apparently moved to strengthen his alliance with Hiram, king of Tyre, and Pekah, king of Israel, around 735, and then unsuccessfully pressured Ahaz (735–716), king of Judah, to join them in their revolt against Assyria.[49] History had shown the Assyrians at the Battle of Qarqar (853) the effectiveness of and danger inherent in the Israelite/Aramean chariotry coalition. Even without Ahaz's plea for help and bribe sent to Tiglath-pileser III (2 Kgs 16:6–8), the Assyrian military strategy necessarily included neutralizing Israel's chariotry forces if it were ever to be successful in defeating Damascus, subduing the Arameans, and securing the trade routes to the Mediterranean coast and Egypt.[50]

Tiglath-pileser III (745–727) invaded Israel proper in 734. With his Assyrian chariot units, he systematically seized and/or destroyed all of Israel's fortresses, horse compounds, and training centers at Dan, Hazor, Kinrot, Rehov, Beth-shean, Jezreel, and most importantly, Megiddo (ca. 732).[51] He cut off Israelite chariot forces previously stationed at Megiddo from coming to the aid of Damascus. The Israelite chariot-training center at Megiddo was converted into an Assyrian administrative headquarters, but the stables were not destroyed and probably served as the central stabling facility for the Assyrian chariotry.[52]

Undoubtedly, the Israelite chariotry dominated the Jezreel Valley until the systemic destruction of the supporting fortresses by the Assyrian forces. Assyria's military strategy did not include allowing other nations to field viable chariotries and armies. We may speculate that the reemergence of Israel's chariotry under Jeroboam II and its subsequent realliance with the Arameans under later kings was, at a minimum, viewed negatively by Assyria, and perhaps even added to their motivation for invasion. Whatever the reason, the capture of the key horse-management facilities at Dan, Jezreel, Beth-shean, and Megiddo destroyed the nerve center of the Israelite army and basically left Israel defenseless.[53] A large, battle-ready chariotry was impossible to recreate without the infrastructure that these cities had provided throughout the previous 100 years.

It is not surprising that the extensive stables and the horse-training facility at Megiddo were left intact and not damaged by the Assyrians, who had immediate

49. Rainey and Notley, *Sacred Bridge*, 228. See 2 Kgs 16:1–5, Isa 7:1–2, 2 Chr 28:5–8.

50. For a detailed discussion of the campaigns of Tiglath-pileser III between 734 and 732 against Israel in the costal plain, Galilee, and Transjordan, see K. L. Younger Jr., "The Deportations of the Israelites," *JBL* 117 (1998) 201–27.

51. See I. Finkelstein, "Tel Rehov and Iron Age Chronology," *Levant* 36 (2004) 181–88; Niemann, "Core Israel," 832–33.

52. For a discussion on Assyrian-controlled Megiddo during the reign of Sargon II, see J. Peersmann, "Assyrian Magiddu: The Town Planning of Stratum III," in *Megiddo III: The 1992–1996 Seasons* (ed. I. Finkelstein, D. Ussishkin, and B. Halpern; 2 vols.; Tel Aviv: Tel Aviv University Press, 2000) 2:524–34.

53. Assyria's campaign to Gaza in the previous years ensured that the Egyptians would not enter the fray to aid Israel; Becking, *The Fall of Samaria*, 11–12.

use for them. It is likely that the Megiddo fortress was abandoned and taken without a battle; otherwise, we would expect the Assyrians to have recorded the capture of Megiddo in some detail or at least to have mentioned it in the annals of Tiglath-pileser III. There is no mention of the capture of Megiddo in either the Assyrian records or the biblical text, although many cities nearby are mentioned.[54] Given that Israel's warhorses were a precious commodity, efforts to evacuate and relocate the horses from Megiddo would have taken main priority and required a significant amount of man power, thus draining the fortresses of adequate military support.

It would be reasonable to assume that, during Tiglath-pileser's invasion, Israel's horses were evacuated to the south, along with their trainers and other key personnel. They could have been relocated to Samaria or other compounds prior to the invasion, or they could have been dispersed to farmers in the countryside for hiding. It is also possible, notwithstanding the strained political relationship with Judah, that the Megiddo chariotry contingent was evacuated to Lachish, where stables, chambered gates, and most importantly a large, flat, limestone-plastered, interior plaza for training were available.[55] However, it is clear that, 12 years later, Samaria still commanded a sizable chariotry unit. As recorded on the Nimrud Prism, Sargon II (722–705) besieged the capital city in 720.[56] Following the capture of Samaria, he apparently took the best officers and formed a new military unit. He claims to have conscripted 200 chariots for his royal contingent and sent the rest of them to Assyria.[57] The immediate incorporation and redeployment of the Israelite chariot units into the Assyrian

54. Megiddo is not mentioned in the list of defeated cities in 2 Kgs 15:29, which states: "In the time of Pekah, king of Israel, Tiglath-pileser, king of Assyria, came and took Ijon, Abel-beth-maacah, Janoah, Kedesh, and Hazor. He took Gilead and Galilee, including all the land of Naphtali, and deported the people to Assyria."

55. See D. Ussishkin, "Area Pal: The Judean Palace-Fort," in *The Renewed Archaeological Excavations at Lachish (1973–1994)* (ed. David Ussishkin; vol. 2; Monograph Series 22; Tel Aviv: Emery and Claire Yass Publications in Archaeology, 2004) 831–34. The main military, logistic, and administrative center of the Judean fortress system was located at Lachish during the late eighth century; Ephraim Stern, *Archaeology of the Land of the Bible*, vol. 2: *The Assyrian, Babylonian, and Persian Periods* (2 vols.; ABRL; New York: Doubleday, 2001) 143. For horse figurines with mounted warriors found at Lachish, see Raz Kettler, "Section B: Clay Figurines," in *The Renewed Archaeological Excavations at Lachish (1973–1994)* (ed. David Ussishkin; vol. 4; Monograph Series 22; Tel Aviv: Emery and Claire Yass Publications in Archaeology, 2004) 2058–77.

56. Sargon II sponsored a large chariotry, as is evident by his construction of new facilities for mustering and training his chariot horses at Dur Sharrukin (Khorsabad). It is estimated that the compound easily held as many as or more than the 3,000-horse capacity at Fort Shalmaneser at Kalkhu (Nimrud) built earlier by Shalmaneser III; Fales, "Preparing for War," 53.

57. This claim is made in the Sargon II Nimrud Prism IV 32–33. The claim that 50 chariots were taken from Samaria is made in the Annals and Display Inscription 24; Becking, *The Fall of Samaria*, 41–42. See also *ANET*, 284–85; S. Dalley, "Foreign Chariotry and Cavalry in the Armies of Tiglath-pileser III and Sargon II," *Iraq* 47 (1985) 31–48. Sargon also took similar action upon conquering Hamath, where he took 300 chariots and 600 mounted men and added them to his royal corps. At Carchemish, he also took 50 chariots and 200 men on horseback to add to his army; see *ANET*, 284–85. It is important to

army indicates that, at a minimum, the horses and soldiers were well trained and battle proficient. Although trained horses can be transferred with ease to other horse professionals, there is a period of time while adjusting to each other that the team might not be "battle-ready." Presumably, the Assyrians needed the Israelite charioteers, not just the horses, because numerous West Semitic names appear as officers in the horse lists.[58]

The Assyrians came back to the Levant in 701 to battle for complete control of Judah and Jerusalem. They first secured all of the enemy horses by attacking and capturing 46 cities in Judah, including the fortress at Lachish that served as the main stables and mustering center for Judah. The maintenance of a horse compound at Lachish seems logical because of the less-forgiving terrain around Jerusalem.[59] One of the primary goals of the Assyrian military strategy was undoubtedly that of disabling, or at least neutralizing, the enemy chariotry and cavalry. This practice seems to be referenced in the taunt of the Assyrian commander, after the Assyrians captured Lachish, to King Hezekiah during the siege of Jerusalem, "I will give you two thousand horses—if you can put riders on them!" (2 Kgs 18:23).[60] The prominence of Lachish as a major military and horse center probably explains why the Assyrians considered its capture worth depicting on their palace walls at Nineveh; it was the nerve center of the Judean army. The depiction of the Lachish Battle on Sennacherib's reliefs shows a chariot being thrown down from the towers.[61] The stables at Lachish were identified by David Ussishkin, the archaeologist who excavated the site, as being capable of housing 150 horses in the immediate area near the chambered gates and interior courtyard.[62] Of course, more horses could be kept in corrals and other areas on the site.

Two important points must be emphasized about the eighth-century Assyrian invasions in the Levant. First, every Assyrian invasion into Israel and Judah involved capturing the war-horses, which was essential in destroying the enemy's military defenses, preventing ambushes, and demoralizing the army. Once the Assyrians neutralized the Israelites and later the Judahites, the valuable Phoenician and Arab trade

---

remember that the horses and chariots captured had probably been left behind to guard the city and king; undoubtedly, many more escaped.

58. See tablet 99 in Dalley and Postgate, *The Tablets from Fort Shalmaneser*, 173. For discussion on the chariot and cavalry units incorporated into the Assyrian army from Hamath and Philistia, see K. L. Younger Jr., "'Give Us Our Daily Bread': Everyday Life for the Israelite Deportees," in *Life and Culture in the Ancient Near East* (ed. R. E. Averbeck et al.; Potomac, MD: CDL, 2003) 269–88.

59. The rocky terrain around mountainous Jerusalem makes the area unsuitable for chariot warfare but poses no particular difficulty for cavalry. See the account of Nehemiah's inspection of Jerusalem on horseback (Neh 2:12).

60. For an interesting literary analysis of this passage, see Ehud Ben Zvi, "Who Wrote the Speech of Rabshakeh and When?" *JBL* 109 (1990) 79–92.

61. D. Ussishkin, *The Conquest of Lachish by Sennacherib* (Tel Aviv: Tel Aviv University, Institute of Archaeology, 1982) 62, 73.

62. Idem, "Assyrian Attack on Lachish," 53–86. See also idem, "Area Pal: The Judean Palace-Fort."

routes came under their control.[63] They also established a major trading center at Gaza in 734, prior to the invasions of Israel and Judah and thus had a convenient locale for importing the prized Kushite horses. The importation of Kushite horses began in earnest during the reign of Tiglath-pileser III and continued through Sargon's reign when the envoys from Gaza brought 24 horses as tribute to Kalhu.[64] Second, as explained by Dalley, "It is true to say that the Levantine states alone provided professional equestrian units for the home army of Assyria under Sargon, from the testimony of the royal inscriptions with the backing from the Horse Lists."[65] Horses and chariots were routinely captured as booty, and since the time of Ashurnasirpal II (884–860), professional soldiers with equestrian expertise from the North Syrian states had been incorporated into the Assyrian army in special "deportee" units. However, under Sargon, only from Samaria, Hamath, and Carchemish were complete battle-ready equestrian units taken into the Assyrian army at large.[66] At a minimum, this suggests the superiority of the equine professionalism and the advanced level of training of the horses from these locales.[67]

## Egyptian Invasions

In 925, nearly 5 years after the reported split of the Israelite Monarchy (931/930), Shoshenq (Shishak) I (945–924) attacked the Southern Kingdom of Rehoboam and the Northern Kingdom of Jeroboam I, assaulting locations in the Beer-sheba Valley, the area of Gibeon, and the Jezreel Valley, including Megiddo, where he erected a victory stele.[68] Shoshenq succeeded in capturing fortified cities in both kingdoms, including Taanach, Beth-shean, and Rehov.[69] Some scholars believe that the inva-

63. Becking, *The Fall of Samaria*, 9.

64. Dalley, "Foreign Chariotry," 44. See K. L. Younger Jr., trans., "A Letter Reporting Matters in Kalaḫ (Kalḫu)," *COS* 3:96:245.

65. Dalley, "Foreign Chariotry," 42.

66. Dalley and Postgate, *Tablets from Fort Shalmaneser*, 37.

67. Dalley, "Foreign Chariotry," 39–42; quotation from p. 42.

68. See Finkelstein, "Campaign of Shoshenq I," 122, 128–29.

69. Notably, chariot fittings were found at many of these fortresses, for example, Beth-shean, Megiddo, Gezer, and Gaza, located on or near the "Ways of Horus," the military highway that connected Egypt to the Levant in the Late Bronze III period. Saddle bosses and yoke terminals made of stone and alabaster were found at Beth-shean; additional materials made of local gypsum may indicate that chariot workshops were also located there; Michael G. Hasel, *Domination and Resistance: Egyptian Military Activity in the Southern Levant, 1300–1185 B.C.* (Leiden: Brill 1998) 6–97, 105.

It is interesting to consider whether Shoshenq revisited the former Egyptian forts on the "Ways of Horus," as depicted on the reliefs of Seti I (1291–1278) at Karnak, in an attempt to establish himself in the mold of the former pharaoh. Further research would be necessary to make these connections. The forts established on the "Ways of Horus" were approximately 25 km apart, about the distance that chariot horses can travel without needing to be changed, about one hour of travel when averaging 18–20 mph. See A. Faust, "The Negev 'Fortresses' in Context: Reexamining the 'Fortress' Phenomenon in Light of General Settlement Processes of the Eleventh–Tenth Centuries B.C.E.," *JAOS* 126 (2006) 135–60.

sions were conducted throughout Shoshenq's 21-year reign, rather than in one major campaign.[70]

In the biblical accounts of Shoshenq's invasion (1 Kgs 14:25–26, 2 Chr 12:2–12), only the Chronicler mentions the Egyptian horses and chariots, placing the chariotry force at 1,200. Although some scholars discount the Chronicles account as anachronistic, it merits consideration because the mention of horses and chariots raises interesting questions about battle tactics and other practical aspects about how the Egyptians could have successfully attacked and conquered the 154 sites they claim to have destroyed.[71]

As a practical matter, it is difficult to imagine that Shoshenq would have attacked *without* a substantial chariotry force, because by then Egypt had at least a 500-year history of engaging in chariot battles. Shoshenq was apparently headquartered on the same stretch of territory in the Delta where Pi-Ramesses had once been located, near Ramesses' royal stabling facility for 460 horses.[72] Theoretically, Shoshenq should have been rebuffed by the superior chariotry forces of Rehoboam and Jeroboam I, presumably inherited from Solomon ("fourteen hundred chariots and twelve thousand horses, which he kept in the chariot cities and also with him in Jerusalem").[73] The failure of Rehoboam and Jeroboam I to defeat Shoshenq may be further evidence that later authors enhanced the memory of Solomon and that, in fact, in the late tenth century, neither Judah nor Israel had yet developed a standing army consisting of a large chariotry.[74]

Second, the importance of chariotry forces in siege operations is typically overlooked in scholarly analysis. The question must be asked, "What prevented the occupants under siege from leaving, either en masse or a few at a time, and simply outrunning the attackers?" Once out of arrow range (maximum 250 yards), they would be free.[75] However, if the enemy had chariot patrols and horsemen when besieging a city, it would not be possible for someone to outrun the horses. Horses and

70. E. Lipiński, *On the Skirts of Canaan in the Iron Age: Historical and Topographical Researches* (OLA 153; Leuven: Peeters, 2006) 100.

71. For a detailed description of Shoshenq's invasion and the locations mentioned on the Bubastite Portal at Karnak, see Rainey and Notley, *Sacred Bridge*, 171, 185–89. See also A. R. Schulman, "Military Organization in Pharaonic Egypt," *CANE*, 1:289–302. For the suggestion that Shoshenq's invasion could have taken place late in the reign of Solomon, see D. B. Redford, *Egypt, Canaan, and Israel in Ancient Times* (Princeton: Princeton University Press, 1992) 315.

72. Ibid., 314–15. It is unknown whether these stables were still in use during Shoshenq's reign; perhaps the ongoing excavations at Qantir (ancient Pi-Ramesses) will clarify the issue.

73. See 1 Kgs 10:26, 2 Chr 9:25.

74. See Cantrell and Finkelstein, "A Kingdom for a Horse," 660, for the suggestion that the references to Solomon's large chariotry is a reminiscence of the situation at Megiddo in the eighth century.

75. R. A. Gabriel and K. S. Metz, *From Sumer to Rome: The Military Capabilities of Ancient Armies* (Contributions in Military Studies 108; Westport, CT: Greenwood, 1991) 70.

chariots logically played a key role in the capture of any fortress by placing the attackers within immediate and constant arrow range of any attempts at escape.[76]

According to the biblical text, the next invasion of the Egyptian chariotry occurred around 896, when Zerah the Kushite marched against Judahite forces with "an army of a thousand thousand [*sic*] and 300 chariots" (2 Chr 14:8–15, NJPSV, LXX). Again, this account is found only in Chronicles. In this story, the Kushites were defeated by King Asa of Judah in the Valley of Zephathah near Mareshah. There is no mention of Judah's possessing chariots—only thousands of well-equipped foot soldiers with spears and bows. Although the huge numbers of soldiers on both sides may be suspect, and the story may be imagined, the lesson is that a small chariotry was not always able to defeat a substantially larger infantry. This may be why chariotry forces continued to grow during the Iron Age; they had to be large enough, with enough reserves, to outlast an infantry. In this story, a mere 300 chariots were not enough to make a difference.

Apparently, it was over 200 years before the Egyptian chariotry entered the Levant again. The Battle of Eltekeh (Tell esh-Shalaf) occurred during Sennacherib's campaign against Judah (701), after he had conquered the cities of Phoenicia and marched into Philistia, taking Joppa, Ashkelon, and Ekron.[77] The Egyptian army traveled into the Levant to support a number of Philistine towns in their rebellion against Assyria. The battle account is recorded on col. III of Sennacherib's Prism (or: the Chicago Prism, ca. 689), which is currently located in the Oriental Institute, Chicago, Illinois.[78]

The two armies met on the plain of Eltekeh, about 6 miles northwest of Ekron and 25 miles north of Gaza. The inscription on Sennacherib's Prism describes how the rulers of Ekron banded together with the Egyptian and Kushite kings, as well as their vast armies of horses and chariots:

> They called out for the kings of the land of Egypt, an army of bowmen, charioteers, and horses of the king of the land of Cush, a host without number; they came to their aid. In the vicinity of the city of Altaqo [Eltekeh] they were arraigned in battle order against me and they were sharpening their weapons.[79]

The Battle of Eltekeh was fought on open terrain. Although it is considered by some authorities to have been a draw, Sennacherib's later conquests place this conclusion in question.[80] After the battle, the Assyrians did not leave the area, but continued to

76. For a detailed discussion of the role of the chariot in protecting a city from siege, see Gabriel, *Military History*, 23.

77. Rainey and Notley, *Sacred Bridge*, 241.

78. Two other sources, The Rassam Cylinder (700) and the Taylor Prism (691), also record these battles. For a detailed account of the Assyrian conquest of Phoenicia and Philistia, see Rainey and Notley, *Sacred Bridge*, 240–41.

79. Ibid., 242.

80. Gallagher, *Sennacherib's Campaign to Judah*, 121.

capture other cities nearby, including Timnah and Azekah.[81] In fact, Sennacherib's success at Eltekeh gave the Assyrians control over the main highway leading up from the south. The defeat of the Egyptian chariotry was significant because it allowed Sennacherib to turn his focus to Judah where, as noted by Naʾaman, the Assyrians fought "tooth and nail" while ravaging the cities, taking Lachish, and putting Jerusalem under siege in 701.[82]

In 609, Pharaoh Necho II (610–595) passed through Megiddo on his campaign north to help the Assyrian king, Ashur-uballit (612–609), regain control of Harran, a significant horse-breeding area, from the Medes and Babylonians, who had captured Nineveh in 612 (2 Kgs 23:29–30, 2 Chr 35:20–24).[83] According to the account in Chronicles, Josiah (640–609), king of Judah, tried to stop the Egyptian advance to aid the Assyrians by staging a chariot battle on the Megiddo plain. The Chronicles account also states that Josiah disguised himself as an ordinary soldier and rode in a common battle chariot, a tactic unsuccessfully tried by Ahab in battle against the Arameans over 200 years earlier (1 Kgs 22:30–35).

Because the 2 Kgs 23:29–30 text does not mention the details of the chariot battle and simply says, "King Josiah marched toward him, but when he confronted him at Megiddo, [Pharaoh Necho] slew him," some scholars believe that Necho simply murdered Josiah and that no battle ensued.[84] Many hypotheses have been suggested regarding the death of Josiah.[85] Both the Kings and Chronicles accounts as well as the dependent accounts in the later works 1 Esd 1:25–32 (Apocrypha) and *Antiquities of the Jews* (Josephus) agree that the confrontation between Josiah and Necho took place at Megiddo and that Josiah was transported by chariot back to Jerusalem for burial.[86]

81. Rainey and Notley, *Sacred Bridge*, 242. N. Naʾaman, "Sennacherib's 'Letter to God' on His Campaign to Judah," *BASOR* 214 (1974) 25–39.

82. Ibid., 146; however, Donald Redford suggests that the Assyrians may have been less than successful at Eltekeh: "There can be no doubt that it was an unexpected and serious reverse for Assyrian arms, and contributed significantly to Sennacherib's permanent withdrawal for the Levant" (*Egypt, Canaan, and Israel*, 353).

83. The role of horses and chariots in the capture and siege of Nineveh is referenced dramatically by the seventh-century prophet Nahum (Nah 2:3–4, 13; 3:2–3).

84. See, for example, N. Naʾaman, "The Kingdom of Judah under Josiah," in *Ancient Israel and Its Neighbors: Interaction and Counteraction* (*Collected Essays*; 3 vols.; Winona Lake, IN: Eisenbrauns, 2005–6) 1:329–98.

85. Naʾaman, for example, suggests that Necho moved his army by sea and "set out on foot" to aid the Assyrians in Syria, via Megiddo, where he met Josiah to require his oath of fealty and killed him when he refused. However, it is difficult to imagine an Egyptian king "setting out on foot" anywhere, given Egypt's 1,000-year history linking the prestige of pharaohs to their battle prowess and chariotry skill. This is especially true because the preceding Kushite Dynasty, as discussed in chap. 2, was so intimately involved with horses and chariotry. Regarding Naʾaman's suggestion that Necho and his army arrived by sea, to me the large-scale shipment of horses by sea at this time seems highly improbable. See ibid., 380–81.

86. Josephus, *Ant.* 10.74–77.

The debate about whether a battle took place at Megiddo in 609 depends largely upon whether it is plausible to believe chariotry forces were available to both Egypt and Judah, as described in Chronicles.

However, the historical context suggests that horses and chariotry were just as important in the late seventh century as they had ever been in the convention of warfare. For example, we know from the Babylonian Chronicles that, in 616, Necho's predecessor, Psammetichus (664–610), who presumably took over the famous Kushite horses and chariotry of the Twenty-Fifth Dynasty discussed above, aided the Assyrians in their battles against the Babylonians in the Upper Euphrates area, all of which involved horses and chariotry.[87] Psammetichus's chariotry had to move through the Levant to reach the battle arena near the Euphrates. He perhaps rested his horses and troops at the Megiddo stable facility, under Assyrian control since 732.[88]

In 606, the Egyptian army headquartered at Carchemish attacked the Babylonians and was successful in causing their withdrawal from the area. This feat could not have been accomplished without an extensive chariotry, given the large Babylonian army with its own substantial equestrian units.[89] In 605, Nebuchadnezzar II, the king of Babylon, battled the Egyptian army at Carchemish and defeated them soundly.[90] In late 605 and 604, the Babylonians invaded the Levant, conquering Ekron and Ashkelon and attacking Gaza.[91] In 601, Nebuchadnezzar returned with his horses and chariots and fought a pitched battle in Egypt, resulting in severe losses to both sides.[92] After this battle, the Babylonian Chronicle reports: "The king of Akkad [Babylon] stayed home (and) *refitted his numerous horses and chariotry*."[93] In 557, Babylonian King Neriglissar orchestrated his army and cavalry in an ambush attack against the king of Pirindu (Syria) that resulted in the capture of the army and "numerous horses."[94] The role of horses and chariots in battle are mentioned in the Babylonian Chronicle over a 55-year period from the fall of Nineveh to the capture of the "numerous horses" in Syria in the mid-sixth century.

87. Babylonian Chronicle 3, lines 10–14 (A. K. Grayson, *Assyrian and Babylonian Chronicles* [Texts from Cuneiform Sources 5; Locust Valley, NY: Augustin, 1975; repr. Winona Lake, IN: Eisenbrauns, 2000] 91–92).

88. Assuming that the entire army averaged 13–15 miles per day, the approximately 100-mile march from Gaza to Megiddo probably took about 8–10 days, the distance that large groups of horses and men could reasonably travel before needing rest. For the Battle of Megiddo, Thutmose III claims to have moved an army of 10,000 from Gaza to Yehem, near Megiddo in 10 days; see Cline, *Battle of Armageddon*, 17.

89. Babylonian Chronicle 4, lines 24–26 (*ABC*, 98).

90. Babylonian Chronicle 5, lines 1–8 (*ABC*, 99).

91. Rainey and Notley, *Sacred Bridge*, 262–63.

92. Babylonian Chronicle 5, lines 5–7 (*ABC*, 100).

93. Babylonian Chronicle 5, line 8 (*ABC*, 101).

94. Babylonian Chronicle 6, lines 1–9 (*ABC*, 103).

Not only does the Babylonian Chronicle attest the importance of the horses in warfare during this period, but several biblical texts reflect the widespread fame and dread of the Babylonian equestrian forces. Concerning the impending Babylonian invasion, the late-seventh-century prophet Jeremiah warns:

> Lo, he ascends like clouds,
> *His chariots are like a whirlwind,*
> *His horses are swifter than eagles.*
> Woe to us, we are ruined! (Jer 4:13)
>
> *At the shout of horseman and bowman*
> The whole city flees.
> They enter the thickets,
> They clamber up the rocks.
> The whole city is deserted,
> Not a man remains there. (Jer 4:29)
>
> Look, an army is coming from the land of the north;
> A great nation is being stirred up from the ends of the earth.
> They are armed with bow and spear;
> They are cruel and show no mercy.
> *They sound like the roaring sea as they ride on their horses;*
> They come like men in battle formation to attack you, O Daughter of Zion.
> We have heard reports about them, and our hands hang limp. (Jer 6:22–25, NIV)
>
> *The snorting of the enemy's horses is heard from Dan;*
> *At the neighing of their stallions the whole land trembles.*
> They have come to devour the land and everything in it,
> The city and all who live there. (Jer 8:16, NIV)

Specifically in reference to the defeat of the Egyptians by the Babylonians at the Battle of Carchemish (605), Jeremiah inveighs:

> Get ready buckler and shield, and move forward to battle!
> *Harness the horses; mount, you horsemen!*
> *Advance, O horses, dash madly, O chariots!*
> Let the warriors go forth. (Jer 46:3–4, 9)

The renown of the Assyrian and Babylonian equestrian warriors is also attested in an early-sixth-century context in Ezek 23:14–24. And also, specifically concerning Nebuchadnezzar of Babylon, "a king of kings, with horses, chariots, and horsemen," in Ezek 26:10–11:

> *His horses will be so many that they will cover you with dust.*
> *Your walls will tremble at the noise of the war horses, wagons and chariots when he enters your gates. . . . The hoofs of his horses will trample all your* streets; he will kill your people with the sword. (NIV)

Additional mention of the dread of the Babylonian horses is noted during the same time period by the prophet Habakkuk:

> I am raising up the Babylonians, that ruthless and impetuous people,
> Who sweep across the whole earth to seize dwelling places not their own.
> They are a feared and dreaded people;
> They are a law unto themselves and promote their own honor.
> *Their horses are swifter than leopards, fiercer than wolves at dusk.*
> *Their cavalry gallops headlong; their horsemen come from afar.*
> They fly like a vulture swooping to devour;
> They all come bent on violence. (Hab 1:6–8)

Given the historical context supporting the extensive use of horses and chariots in battle during the Babylonian war against the Assyrians and Egyptians in the late seventh and early sixth centuries, it seems irrational to think that Necho and Josiah would have met each other at Megiddo *without* their horses and chariots. Not only was the vast Megiddo plain one of the most suitable places for chariot battle in the world, its strategic location on the route to the north made it the most logical place to deter the Egyptians, if that was Josiah's intention.

The Babylonian revolt against Assyria began in 626, and at some point shortly thereafter caused Assyria to reduce its presence in the Levant. It is generally agreed that Megiddo had ceased to be an Assyrian administrative center sometime after the death of Ashurbanipal in 631; however, the stables and training area remained functional for horses. They could have been used by Egypt during its trips north to assist Assyria. Or, if Judah had taken control of Megiddo, the stables could have housed the Judean chariotry. Even if we assume that the Egyptians had filled the Assyrian vacuum and used Megiddo for a convenient way station in 616, after that, their main army was occupied near the Euphrates aiding the Assyrians. If Josiah decided to rebuild Judah's chariotry through an aggressive breeding program, presumably both Assyria and Egypt would have been too busy fighting Babylon to stop him. Alternatively, if Egypt considered Judah a vassal during this period, it may have encouraged the buildup of a Judean chariotry at Megiddo that presumably would have been accessible to Egypt and Assyria. Either way, stationing a chariotry at Megiddo seems quite plausible. The distance from Jerusalem to Megiddo is only 60 miles and could be traveled by horseback or in chariot in less than a day. Mustering an army there from Judah could be accomplished easily. Given Megiddo's already-lengthy history as a battlefield and horse-training center, it was a convenient and practical place for a battle.[95]

Perhaps Josiah planned to ambush Necho as he brought troops through the Musmus Pass. Or, more likely, Josiah knew the size of Necho's army (because it had been

95. For a discussion of Josiah's strategic options in staging a battle at Megiddo, see Cline, *Battle of Armageddon*, 90–100.

on the move from Gaza for days prior to arriving at Megiddo) and apparently thought that he had enough chariotry and infantry available to challenge the Egyptians in an open, pitched battle. This may indicate that Egypt was advancing with a contingent of reinforcements to supplement the army that it had left stationed at Carchemish with the Assyrian troops. If so, both armies may have been relatively small by past battle standards.[96] The story about Josiah disguising himself to avoid detection may represent a commonly accepted battle tactic to confuse the enemy. All sources agree that Josiah died at Megiddo. I believe that the evidence—admittedly contextual and circumstantial—best supports the notion that Josiah died in his chariot at the outset of the battle. The battle at Megiddo in 609 was the last recorded chariot battle in Israel and may have been one of the last great (albeit, brief) chariot battles in history resulting in the death of a king.

## *Invasion of Moab (2 Kings 3)*

Israel's (and Judah's) involvement in Iron Age warfare was not solely for defense purposes. The biblical text records that shortly after the Battle of Ramoth-gilead the kings of Israel, Judah, and Edom joined forces to invade and punish Moab (ca. 850). The advantage of proximity to reinforcements and short distances between fortresses was lost during the allied invasion of Moab, because the kings routed their armies the long way around the southern tip of the Dead Sea through the desert of Edom (2 Kgs 3:9).[97] Had the coalition launched the campaign from Jerusalem or from the northern tip of the Dead Sea, the distance to the Moabite border was only 20 miles; the capital at Dibon was 20 miles farther. The allied army could have made the direct, 40-mile trek in one or two days, albeit over somewhat more difficult terrain and with threats of attack from the Arameans or Ammonites. By choosing the southern route through the Edomite desert, they doubled their distance.

According to the biblical text, en route to the invasion the troops became lost and spent seven days wandering in the desert before running out of water for themselves and their horses. Although this perilous situation was remedied by the miraculous intervention of God answering Elisha's petition for water, the launch of an invasion without adequate water supplies suggests that the combined forces' prior battle experience involved short distances and brief campaigns.

96. However, Egyptologist Donald Redford suggests that Necho "spent the winter mustering a much larger expeditionary force than had heretofore been sent into Asia" ("*Egypt, Canaan, and Israel,*" 448).

97. It also points out the dangers and increased risks facing an invading army on desert expeditions; see Rainey and Notley, *Sacred Bridge*, 205.

# Chapter 7

# *From Chariotry to Mounted Combat*

*They sound like the roaring sea as they ride on their horses;*
*they come like men in battle formation to attack you,*
*O Daughter of Zion.*
*(Jer 6:23, NIV)*

As early as the seventeenth century, with the invention of the composite bow and penetrating bronze arrowheads, which allowed rapid fire at a distance, the war chariot had become the most lethal weapon in existence.[1] Bronze Age battles were almost always chariot-centered as attested in the accounts of the Battle of Megiddo (1479) and the Battle of Qadesh (1285). The Greeks ceased using chariots in war after the Mycenean period (ca. 1200).[2] However, Alexander's cavalry forces faced Persian chariots at the Battle of Issus (333) and at the Battle of Gaugamela (331), although they did not prove effective.[3] The Romans, who used chariots only for racing, were surprised to find the British fielding war chariots against them when Caesar invaded Britain in the first century.[4] The change from chariotry to cavalry as the preferred battle convention seems to have been gradual in every culture.

Mounted riders were a part of the Assyrian army as early as Ashurnasirpal II (883–89) and are primarily pictured as archers, riding in pairs. However, by the time of Tiglath-pileser III (745–727) the cavalry units wore armor and carried long spears.[5] This exhibition of superior horsemanship by the mounted rider is a pivotal point

1. The Bronze Age Egyptians are credited with first using archers in chariots, and the Iron Age Assyrians were the first army to deploy the horse-borne archer; see R. A. Gabriel and K. S. Metz, *From Sumer to Rome: The Military Capabilities of Ancient Armies* (Contributions in Military Studies 108. Westport, CT: Greenwood, 1991) 69.

2. For a detailed discussion of chariot warfare in the Late Bronze Age, see R. Drews, *The End of the Bronze Age* (Princeton: Princeton University Press, 1993) 104–34. See also Jorrit M. Kelder, "Horseback Riding and Cavalry in Mycenaean Greece," *Ancient West and East* (2012) forthcoming.

3. See Victor Davis Hanson, *The Wars of the Ancient Greeks* (ed. John Keegan; London: Cassell, 1999) 32–35, 141, 175.

4. J. K. Anderson, "Greek Chariot-Borne and Mounted Infantry," *AJA* 79/3 (1975) 175–87.

5. J. Ellis, *Cavalry: The History of Mounted Warfare* (South Yorkshire: Pen and Sword, 2004) 15.

in history. Mounted warriors, unlike the disadvantaged infantryman, could ride in quickly and close to chariot teams, cut reins, slash the horses' legs, or deliver an incapacitating blow between the eyes, and then flee unharmed. It led to a shift in the most lethal weapon of warfare (and, therefore, the convention) from chariotry to mounted combat. In effect, the superior killing ability of the mounted horseman made chariotry obsolete. Much of this transition took place in Israel and Judah by the end of the Iron Age.

In the Iron Age, with nearly 1,000 years of experience in building chariots, harnessing horses, and conducting warfare, some of the early problems of chariotry were solved, their use streamlined. For example, the Middle Bronze Age chariot, a heavy, four-wheeled battle wagon drawn by onagers wearing nose rings, gave way to the Late Bronze Age chariot, lightweight, with two spoked wheels and drawn by horses wearing bits. During the Iron Age, chariots appear to have been somewhat more resilient, and new bitting techniques, such as the bronze jointed snaffle bit improved the control of horses for both chariotry and mounted riders.

However, chariot warfare slowly became obsolete during the Iron Age as a general revolution in horsemanship occurred.[6] Despite the exceptional capabilities of the chariot, the ability of mounted horsemen to neutralize the chariotry eventually proved so effective that nations had no choice but to replace entire chariot divisions with cavalrymen. Nonetheless, a most intriguing historical question is: why did cavalry supersede chariotry? Three major reasons that merit investigation are improved riding techniques, smaller and faster horses, and new tactics in warfare.

Some portion of the cataclysmic change from chariotry to cavalry in conventional warfare undoubtedly had to do with the breed of the horse used. Egyptian chariot horses were large and thick, with wide backs (ca. one meter), a perfect muscular structure for pulling chariots. However, it was somewhat difficult to sit on them when riding because, by necessity, the rider's legs were widely straddled, with bent knees.[7] This position required a second, nearby rider to hold the reins while the first steadied himself to shoot his arrow.[8] A thinner, smaller horse, such as the Caspian horse used by the Medes, had a narrower back and allowed the rider a longer, straighter leg to

6. For a detailed discussion of improvements in bitting, saddlery, and riding skill as well as the impact of these improvements on warfare, see R. Drews, *Early Riders: The Beginnings of Mounted Warfare in Asia and Europe* (New York: Routledge, 2004) 65–95.

7. This determination is based on the author's assessment and measurements of similarly sized modern horses to the Assyrian horses depicted in the reliefs, presumably the highly favored Kushite breed. See D. Ussishkin, *Conquest of Lachish by Sennacherib* (Tel Aviv: Tel Aviv University Institute of Archaeology, 1982) 88–91, figs. 71–72.

8. An oblique reference to this battle practice may be found in the words of Jehu to Bidkar, his army officer: "Remember how you and I were *riding side by side* behind his father Ahab" (2 Kgs 9:25, emphasis added). See pictures in Y. Yadin, *Art of Warfare in Biblical Lands in Light of Archaeological Study* (trans. M. Pearlman; 2 vols.; New York: McGraw-Hill, 1963) 2:384–85.

clamp around the horse.[9] The narrow muscular structure enabled the rider to control his body more effectively and to communicate commands to the horse better via his legs. This smaller horse, found mainly in the Zagros Mountains, allowed the warrior faster speed and far more maneuverability than the large Egyptian chariot horses.

The advent of bronze and iron bits made riding safer and allowed the rider more control of the horse. Riders also improved their riding style by moving closer to the withers and learning how to "become one with the horse."[10] As insightfully expressed by Drews,

> In the revolution in horsemanship that occurred early in the first millennium B.C. the most important change must have been in the relationship of rider and horse: the riders were now in control, and the horses obeyed.[11]

Ultimately, riders gained experience in riding horses symbiotically (at one with the horse); this advancement in horsemanship allowed for a more skillful use of weaponry.[12]

It is important to note that good riding does not require a saddle or stirrups.[13] In fact, riders probably had a more secure seat without them. The Assyrian reliefs show the cavalrymen sitting on saddle pads, probably made of leather. The saddle pad protected the rider from the horse's sweat, which could have made their backs slippery. Assuming their breeches were also crafted of leather, the leather-on-leather connection with the saddle pad provided some adhesive traction and security for the leg and seat.[14] The saddle pads were held on the horse by a girth, which was in use as early as 2000 B.C.E.[15]

9. For an interesting discussion of the rediscovery of the ancient Caspian horse and its importation to the United States in 1995 C.E., see S. Biggs, "Ancient Treasure," *Horse Illustrated* (Oct. 2002) 66–74. The Caspian is similar to the Arabian in appearance but smaller, typically standing only 10 to 12 hands.

10. Drews, *Early Riders*, 86. For a brief history of the development of riding techniques in Mesopotamia, see M. Weszeli, "Reiten. A. In Mesopotamien," *RlA* 11:303-7.

11. Drews, *Early Riders*, 87.

12. This same ability to ride symbiotically was also achieved in modern times by the Native American Indians (1680–1870 C.E.), who used bow and arrow to hunt buffalo and fight on horseback with no saddle or stirrups. They have yet to be surpassed in their horsemanship skills; G. P. Horse Capture and E. Her Many Horses, eds., *A Song for the Horse Nation: Horses in Native American Cultures* (Golden, CO: Fulcrum, 2006) 36–37. For the suggestion that bareback riding was an advanced stage of horsemanship used primarily during scalping raids, see Drews, *Early Riders*, 40. However, this suggestion is somewhat at odds with modern training practice, in which beginning riders are frequently required to ride without stirrups and even bareback to develop a firm seat. Only if one can ride without stirrups successfully, can one ride with them properly, in this author's opinion.

13. Stirrups appeared in India as early as the second century B.C.E. Saddles with horns to secure the legs were in common use by the Roman and Parthian military in the first century C.E.; A. Hyland, *Horse in the Ancient World* (Westport, CT: Praeger, 2003) 52–54.

14. J. E. Curtis and J. E. Reade, eds., *Art and Empire: Treasures from Assyria in the British Museum* (London: British Museum, 1995) 68, pl. 17.

15. Drews, *Early Riders*, 48.

Interestingly, the shift from chariot warfare to mounted combat can be traced in the biblical text. References to ninth- and eighth-century chariot warfare feature prominently in the prophetic rhetoric of Isaiah. For example, eighth-century references to the Assyrian, Egyptian, and Elamite armies include chariots: "When he sees chariots with teams of horses, riders on donkeys or riders on camels, let him be alert, fully" (Isa 21:7); and "Elam takes up the quiver, with her charioteers and horses; Kir uncovers the shield. Your choicest valleys are full of chariots, *and horsemen are posted at the city gates*" (Isa 22:6–7).

By comparison, references from the late-seventh-century and early-sixth-century prophets focus more on mounted horsemen than on chariots: "at the sound of horsemen and archers every town takes to flight" (Jer 4:29); "they sound like the roaring sea as they *ride on their horses*; they come like men in battle formation" (Jer 50:42); and "you will come from your place in the far north, you and many nations with you, all of them *riding on horses*, a great horde, a mighty army" (Ezek 38:15); "their *horses are swifter than leopards*, fiercer than wolves at dusk. *Their cavalry gallops headlong*; their horsemen come from afar" (Hab 1:8, NIV, emphasis added). These passages chronicle the arrival of the ominous mounted warriors from the north, and as mentioned above, may refer to the Babylonian cavalry, which invaded Judah and conquered Jerusalem in 586.

Furthermore, passages regarding the return of the first Israelite exiles from Babylon in 458–455 indicate that they were accompanied by a Persian cavalry contingent for protection and that travel was by horseback rather than in chariots (Ezra 8:22, Neh 1:9). By the composition of the postexilic book of Esther (ca. 460 or later), set in the Persian court, horseback riding had superseded chariotry for ceremonial purposes as well. When the king inquires what should be done for a man whom the king desires to honor, it is suggested:

> For the man whom the king desires to honor, let royal garb which the king has worn be brought, and a horse on which the king has ridden and on whose head a royal diadem has been set; and let the attire and the horse be put in the charge of one of the king's noble courtiers. *And let the man whom the king desires to honor be attired and paraded on the horse through the city square*, while they proclaim before him: This is what is done for the man whom the king desires to honor! (Esth 6:8–9, emphasis added)

By contrast, in the Joseph story, set in a Bronze Age context, Pharaoh bestows highest honors by staging a ceremonial chariot procession:

> Pharaoh further said to Joseph, "See, I put you in charge of all the land of Egypt." And removing his signet ring from his hand, Pharaoh put it on Joseph's hand; and he had him dressed in robes of fine linen, and put a gold chain about his neck. *He had him ride in the chariot of his second-in-command and they cried before him, "Abrek!"* (Gen 41:41–43, emphasis added)

Although the biblical texts are notoriously difficult to date with certainty, it is not unreasonable to assume, in general, that traditions that incorporate chariotry may be historically more distant in time than traditions with horseback riding.

Some of the postexilic prophets in the biblical text reference both chariots and mounted horsemen as effective components of warfare. For example, "I will take away the *chariots from Ephraim and the war-horses from Jerusalem* and the battle bow will be broken" (Zech 9:10, NIV); "I will *overturn chariots and their drivers. Horses and their riders will fall*, each by the sword of his fellow" (Hag 2:22); and "They have the *appearance of horses; they gallop along like cavalry. With a noise like that of chariots* they leap over mountaintops . . . like a mighty army drawn up for battle" (Joel 2:4–5, NIV). These passages reflect the memory that both chariots and mounted warriors were involved in battle during the Iron Age.

In addition, the "Azekah" Inscription describes Sennacherib's battle with the kings of Philistia, mentioned above, and his invasion into Judah ca. 701 is also found in the Azekah Inscription.[16] It is interesting that this inscription mentions no chariotry but instead states that the city of Azekah had "seen [the approach of my cav]alry and were afraid."[17] By contrast, Sennacherib's battle against Babylon and Elam ca. 689, recorded on the Bavian Rock Inscription, describes chariotry forces at work:

> (As for) the king of Elam and the king of Babylon, the fear of my mighty battle overcame them and they defecated in their chariots and they fled back to their lands in order to save their lives.[18]

The Azekah Inscription contains one of the few references to mounted riders in Judah, a nation whose kings are typically portrayed in the biblical text as riding in chariots rather than on horseback.[19] A passage from Isaiah suggests that the notion of fleeing the enemy on horseback was well known to the Judahites in the late eighth century:

> You said, "No, we will flee on horses." Therefore, you will flee! You said, "We will ride off on swift horses." Therefore your pursuers will be swift! (Isa 30:16, NIV)

Despite this practicality, the Egyptian army, according to the perception of the biblical text appeared to remain steadfast in its dedication to chariotry during the eighth and seventh centuries:

> Woe to them who go down to Egypt for help, who trust in horses and chariots, for they are many; and in horses, which are a great multitude; and have not trusted in the Holy One of Israel, and have not sought the Lord. (Isa 31:1, LXX)

16. M. Cogan, trans., "Sennacherib: The 'Azekah' Inscription," *COS*, 2.119D:304–5.

17. Na'aman, "Sennacherib's 'Letter to God,'" 136.

18. Cogan, trans., "Sennacherib: The 'Azekah' Inscription," *COS*, 2:119D:305.

19. There are also a few references to mounted riders in Israel: messengers meeting Jehu (2 Kgs 9:16–28); Ben-hadad, king of Aram, escaping on horseback from Ahab (2 Kgs 20:20); and numerous references to invading Babylonians on horseback (Jer 4:29, 6:23; Ezek 38:15).

It is evident that from the inscription and the biblical text that, in the late eighth century, both mounted riders and chariotry played important roles in combat. Eventually, mounted combat replaced chariotry in conventional warfare, but during the Iron Age chariots were significant as the preferred mode of transportation for kings and for their usefulness in battle.

The graceful "prancing horse" seal impressions on large storage jars found in Jerusalem, En-Gedi, and other sites in the Shephelah dating to the late eighth century (see cover) display a familiarity with the anatomy of horses and fine artistic craftsmanship. These seal impressions on storage-jar handles throughout Judah may indicate a stamp used by an important official of the royal administration.[20] Perhaps the storage jars contained grain for horses in training. This type of elegant prancing-horse motif is unknown elsewhere in antiquity until somewhat similar horses appear on Greek coinage in the fourth century.[21]

20. Gabriel Barkay, "'The Prancing Horse': An Official Seal Impression from Judah of the 8th Century B.C.E.," *TA* 19 (1992) 124–29.

21. For an insightful review of Greek cavalry and equestrian matters, including a detailed explanation of the superior skills of bareback riders in warfare, see Robert E. Gaebel, *Cavalry Operations in the Ancient Greek World* (Norman: University of Oklahoma Press, 2002) 165.

## Chapter 8

# *Conclusion*

*At your rebuke, O God of Jacob,*
*both horse and chariot lie still.*
*(Ps 76:6, NIV)*

In our nuclear age of sophisticated electronics and high-tech weaponry, it is easy to lose the sense of the astounding power of the horse that people in the ancient world witnessed routinely. Unfortunately in modern times, false notions about the horse and riding have infiltrated historical analysis—for example, that the lack of stirrups and saddles prevented superior horsemanship in battle, or that horses required stables to survive and were prohibitively expensive to maintain, or that Israel was unsuited for raising and training horses. In fact, the archaeological and epigraphical evidence as well as basic equestrian knowledge prove the contrary. The horse was key to the survival of nations, and every country, including Israel and Judah, did whatever was necessary to support and train its war-horses and equestrian warriors.

Iron Age architects in both Israel and Judah designed fortresses with horse-care management as a priority. Cavalry and chariotry developed and thrived under especially advantageous conditions: favorable terrain, short distances between fortresses, adequate pasturage and water, adequate stables, convenient hitching chambers, and the expansive training centers. The military headquarters and equestrian center at Lachish, the horse compound at Jezreel, and most importantly the elaborate stables / training center at Megiddo were essential to the effective management of the armies' most valuable weapon, the horses. Archaeological evidence in both Israel and Judah supports the presence of an active chariotry and horse-management system during the Iron Age.

Battles in Iron Age Israel occurred relatively near the Israelite fortresses, unlike conflicts staged in Egypt and Assyria for which weeks of travel were necessary to reach the battlefield. In most cases, Israelite horses could be brought back to Megiddo, the Jezreel compound, or any other nearby fortress for rest on a daily basis. Other than at the Battle of Qarqar and perhaps the invasion into Moab (2 Kgs 3:9), the horses did not have to endure campsite conditions for an extended period of time, as did the horses of the invading armies. Israel was perfectly suited to the development of an extensive defensive network based on chariotry and cavalry.

It is remarkable that, after the fall of Samaria, the Assyrians redeployed a large Israelite chariot force to fight as a unit of the royal army on its northern borders. It is also significant that the king of Assyria selected an Israelite equestrian to train the crown princes and designated many Samarian charioteers as officers in the Assyrian army. This speaks volumes about the level of equestrian expertise in Israel.

The biblical text, perhaps more explicitly and extensively than any other ancient record chronicles the importance of the war-horse and the power it evoked in the imagination of the people. Thirty of the 39 books in the Hebrew Bible mention horses, chariots, or mounted warriors in some context, usually military, but also poetic and ceremonial. The biblical literature documenting horses and their role in the culture of Israel and Judah covers a historical period of almost 400 years. Whether considering the bravery of the war-horse so eloquently described in Job, the daring chariots of the heavenly horsemen of Israel in 2 Kings, or the fear and dread of the invading Babylonian cavalry in Jeremiah, the various biblical authors regard the power of the war-horse with an awe approaching reverence.

At any given time during the ninth and eighth centuries, horses were the single most-important force standing between Israel and extinction. If Israel had been a nation of only a few hundred horses, it would have been easily and quickly overwhelmed by its mighty neighbors, who possessed brilliant armies with thousands of horses. Or worse, it could have collapsed into one large battlefield, as Aram, Egypt, and Assyria fought for expanded borders and control of the valuable trade routes.

Both Israel and Judah developed and maintained statehood in large part due to their ability to ward off the enemy armies with their own effective chariotries and mounted warriors. From a strictly historical perspective, without the infamous "horsemen of Israel," Israel and Judah might never have existed long enough to be remembered. Undoubtedly, the "horsemen of Israel" have earned the right to prominence in the Monarchic history of both nations.

# Index of Scripture

## Old Testament

## New Testament



## Apocrypha

# Index of Authors